Hamlyn all-colour paperbacks

Electricity

D. R. G. Melville

illustrated by
Whitecroft Designs Limited

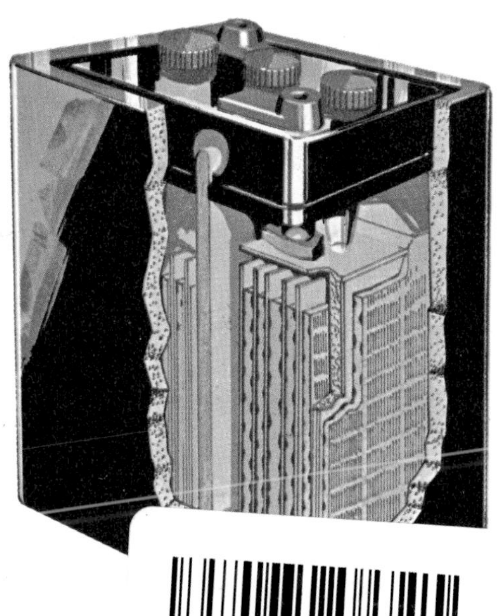

Hamlyn
London · New York · Sydney · Toronto

FOREWORD

The purpose of this book is to explain what electricity is, its manifestations and how it behaves, and then to show how it is produced, distributed and used in the modern world. Finally, a brief look will be taken at possible future developments of the use of electricity.

Treatment of such a vast subject as electricity, in a book of this size, must perforce be brief, since complete books have been written on topics covered in one page of this one, but it is hoped that readers will obtain an understanding of the broad outlines of electrical theory and usage. For those who are interested in more detailed explanations, a bibliography of selected references for future reading is included on page 156.

D.R.G.M.

Published by the Hamlyn Publishing Group Limited
London · New York · Sydney · Toronto
Astronaut House, Feltham, Middlesex, England

Reprinted 1974, 1976
ISBN 0 600 00121 0

Photoset by BAS Printers Limited, Wallop, Hampshire
Colour separations by Schwitter Limited, Zurich
Printed in Spain by Mateu Cromo, Madrid

CONTENTS

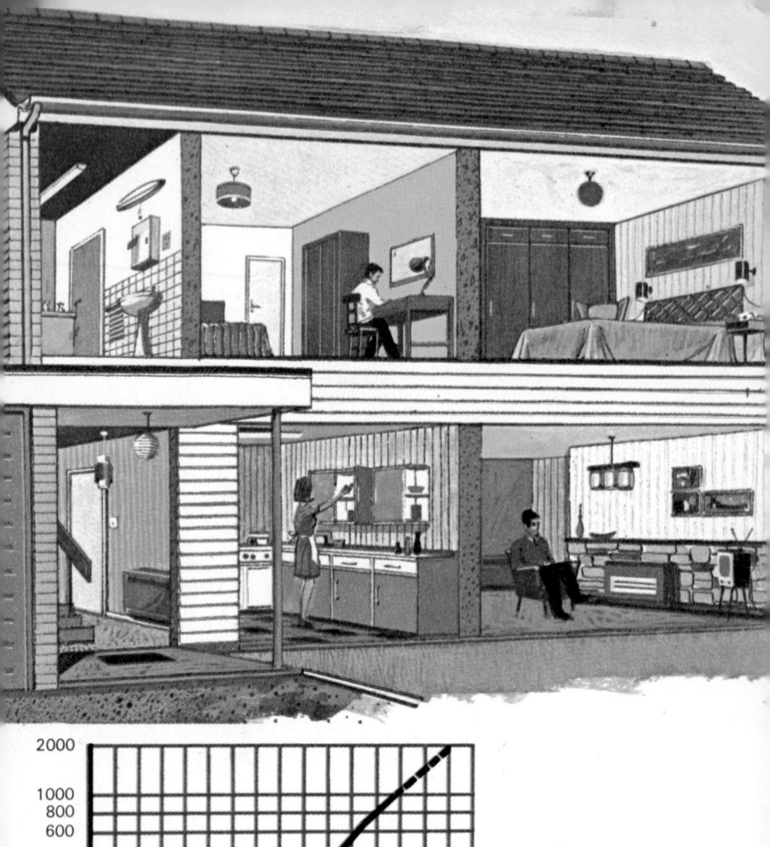

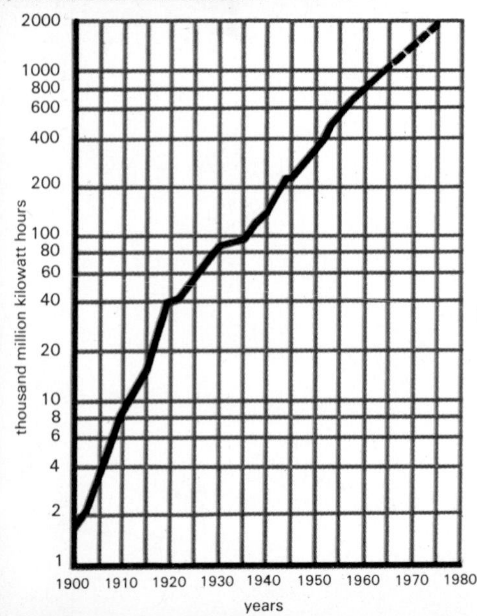

Trends in consumption of electricity

INTRODUCTION

Electricity is probably the 'eighth wonder of the modern world'. Its discovery and exploitation have revolutionized the homes and industries of the civilized world during the last century. Life today is utterly dependent on it. Look at the effects of a power failure lasting only a few hours on a city, or even part of one. Think of all the devices that are based on it; lights, radio, television, fires, motors to drive the wheels of industry. The list is unending.

Electricity is one of the forms of energy (the power to do work) and it can be produced from other types of energy, such as the energy of a chemical reaction or of mechanical rotation, to name but two. Its great advantages over all other types of energy are its cleanliness (absence of smoke or smell), flexibility and efficiency. Electricity is also much easier to transport over long distances than other forms of energy, requiring only two wires, not much over an inch in diameter, as opposed to large pipelines for oil and gas or large containers for coal. The use of electricity has been growing rapidly over the years and the demand is now doubling every seven to ten years.

Along with electricity we must also consider magnetism. The flow of an electric current sets up a magnetic field and if a magnet moves near a wire, a current will flow in the wire.

Terminology and prefixes

The commonest shorthand methods of expressing very large and very small numbers are set out below.

Magnitude	Numerical value	Abbreviation	Prefix	Symbol
Million	1,000,000	10^6	Mega	M
Thousand	1,000	10^3	Kilo	k
One thousandth	0·001	10^{-3}	Milli	m
One millionth	0·000001	10^{-6}	Micro	μ

For example, one megavolt or 10^6 volts is 1,000,000 volts and one milliampere or 10^{-3} amperes is 0·001 amperes.

Historical outline

Although unconscious of the fact, the earliest users of electricity were the cavemen. Trees set alight by lightning strokes were probably their first source of fire.

The first recorded production of electricity was in 600 BC in Greece. After rubbing with a cloth, the mineral amber was found to attract fluff and loose hair, but if subsequently wetted, the amber lost its attractive powers. This was the discovery of static electricity and is acknowledged by the fact that the word electricity is derived from the Greek word for amber, *elektron*. However, at the time this phenomenon was considered to be a novelty and not of much importance.

Magnetism, also discovered early, became of much more importance. The first known magnets were a natural mineral, an iron oxide known as *lodestone*. They came from a town in Asia called *Magnesia*. In 1300 BC records existed of 'attracting rings' being a part of even earlier religions. Legend has it that the Chinese had a 'south pointing chariot' about 1000 BC for use on caravan

journeys but the first documented use of the compass was in AD 700, again in China. Evidence to date implies that the compass was discovered independently in a number of countries and was first used on a large scale in the late twelfth century.

Gilbert, in 1600, carried out the first extensive investigation of the phenomenon of magnetism. He thought that magnetic substances gave out an 'effluvium' which could not be seen or felt but, like an odour, continued for years without any shrinkage of the object. The effluvium drew any substance it contacted back towards the material that it had emanated from.

For the next 150 years many people 'played' with static electricity, producing larger and larger amounts of it by mechanizing the rubbing process. Electric sparks were discovered but no serious advances were made, since no method of storing the static electricity produced had yet been invented.

At the University of Leyden, Professor Muschenbrock discovered, by accident, that static electricity could be stored in what came to be known as the Leyden jar. It consists of a water-filled glass jar encircled by a metal band. The electricity is stored in the glass between the metal and the water.

By this time two types of static electricity had been noted. Each type attracted bodies of neutral or opposite electrification but repelled ones of similar electrification. Franklin proposed the important concept that there is only one type of electricity but some objects have an excess of it and others a dearth. However, Franklin is most famous for showing

(*Left*) Franklin's kite experiment
(*Below*) Simple voltaic cell

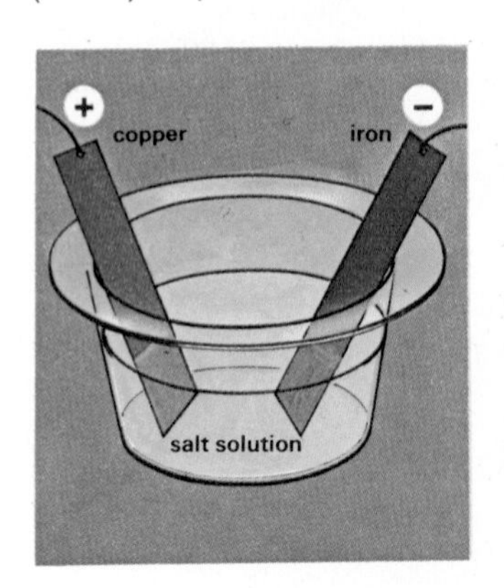

that lightning is a form of static electricity, akin to the electric spark. By flying a kite with a metal cord in a storm he charged a Leyden jar. He was lucky not to be killed.

The Leyden jar could only provide small amounts of electricity before it needed replenishing and it was the Italian, Volta, who in 1798 produced electricity, chemically, by placing copper and iron rods in a glass jar filled with a salt solution. Voltaic piles, the predecessors of the modern battery, were the first sources of continuous electricity.

This led to great advances in the understanding of electricity. Oersted showed what had long been suspected: that an electric current had a magnetic field. When he passed a current from a voltaic pile through a wire suspended over a compass, the needle deflected. Following this Ampère discovered that currents flowing in two parallel wires would attract or repel, depending on the direction of current flow.

Ampère distinguished between voltage (potential) and current, and devised the galvanometer for measuring the latter. This forerunner of the modern ammeter consisted of a coil of wire wrapped around a compass, the deflection angle being proportional to the amount of current flowing.

However, it was George Ohm who in 1827 discovered the fundamental relationship of electrical science, that between voltage and current. Ohm's law states that in any given circuit the voltage divided by the current is a constant.

Force between conductors

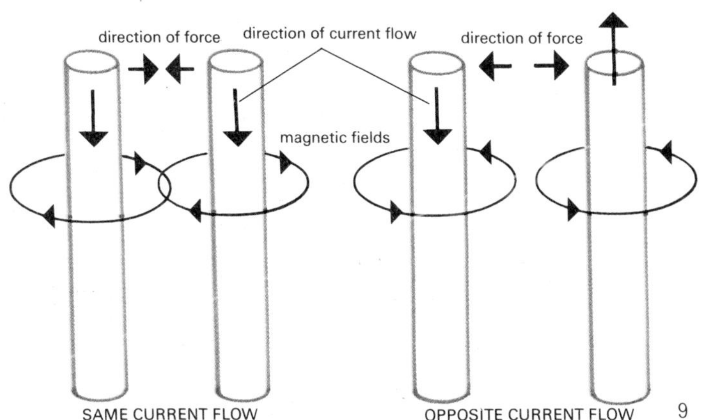

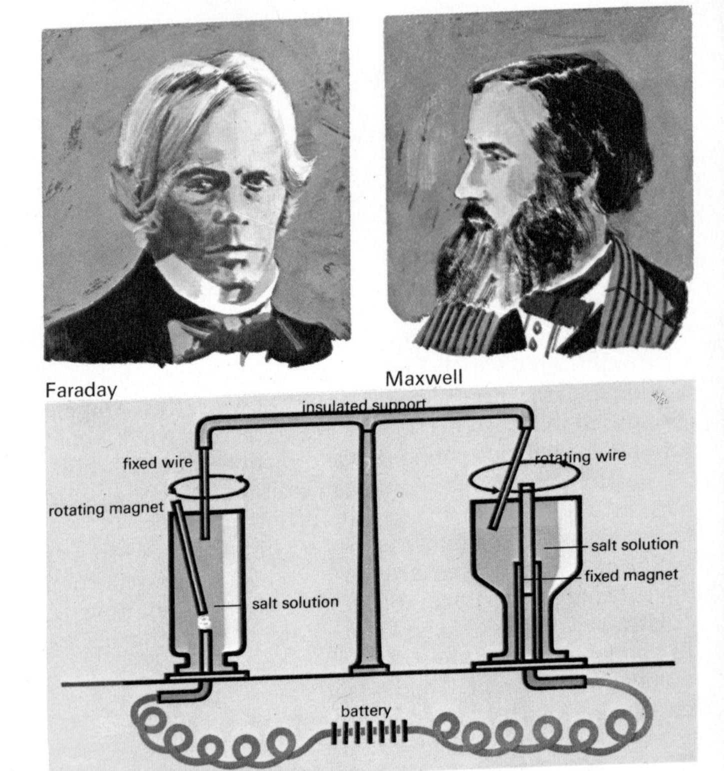

Faraday's demonstration

In 1821 Faraday first demonstrated electromagnetic rotation, with a suspended magnet rotating around a fixed wire and a wire rotating around a fixed magnet. His greatest discovery was, however, that electricity could be produced magnetically, as had long been suspected, but that the magnetic field must be continually changing in magnitude to so do. Faraday set up two adjacent but unconnected coils of wire, with a battery and switch in one and a galvanometer in the other. Only on opening or closing the switch, when the magnetic field due to the current was increasing or decreasing, did the galvanometer needle move. This first transformer led to the use of electricity as we know it today.

Maxwell expressed Faraday's discoveries in mathematical form. He set out the relationship between electricity and magnetism and postulated, and later proved experimentally, that light is a form of electromagnetic radiation. His theories predicted that electric waves could be propagated through air like ripples on the surface of a pond and Hertz, in 1888, proved this experimentally, thus laying the foundations for radio and television broadcasting.

At this time many scientists were investigating the fundamental nature of electricity. Crookes found that an electric field caused an evacuated glass tube to glow and that when all the air was removed, the glass walls glowed green-yellow, when struck by invisible rays (called cathode rays since they emanated from the negative electrode). These rays could be deflected by magnetic fields or stopped by metal screens.

By measuring the deflection of these cathode rays in magnetic and electric fields, J. J. Thomson calculated their mass to electric charge ratio. By measuring the electric charge in a separate experiment, it was found that the mass of the particle was $\frac{1}{1836}$ of the hydrogen atom. This was the first discovery of the electron and led to the important proposal that the atom was not indivisible but contained particles having electric charges. Subsequent measurements of positive rays from the anode indicated their mass to be $\frac{1835}{1836}$ of a hydrogen atom. This led directly to an understanding of the construction of the atom and the nature of electricity and finally led to the atomic and hydrogen bombs (see *Atomic Energy*, a companion volume to this book).

Thomson's experiment

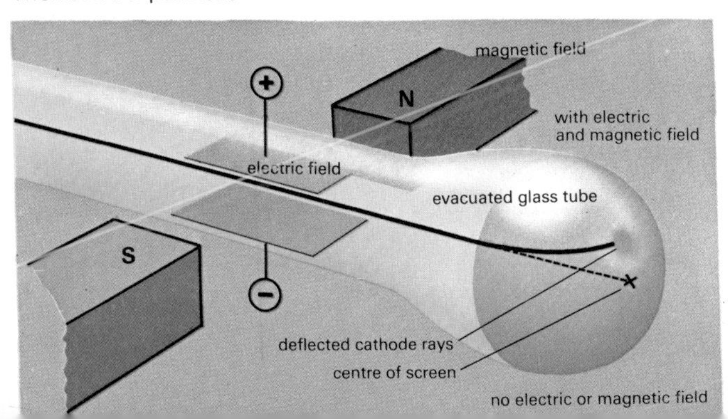

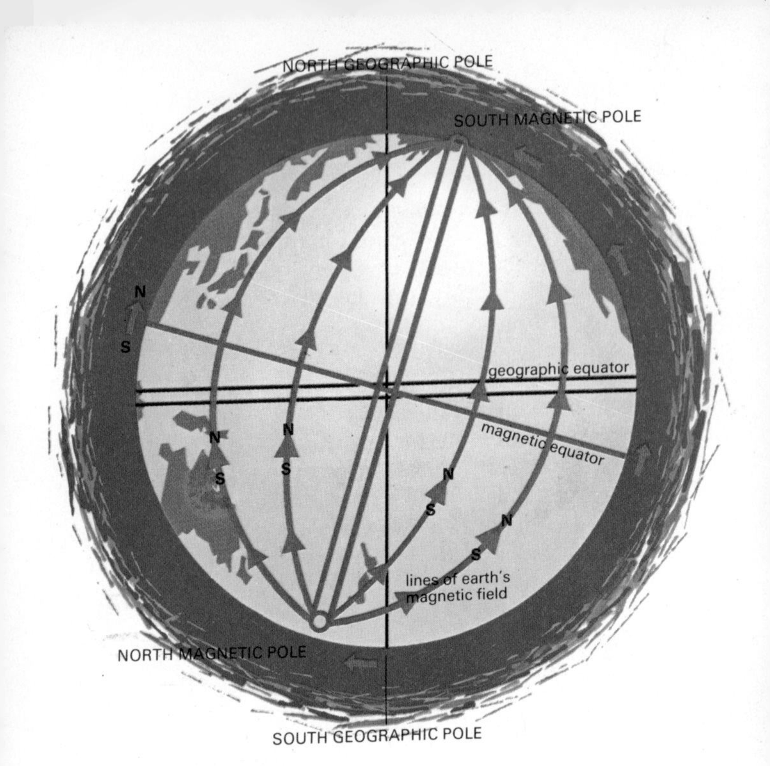

(*Above*) Magnetic field of the Earth. (*Below*) Magnetic interaction

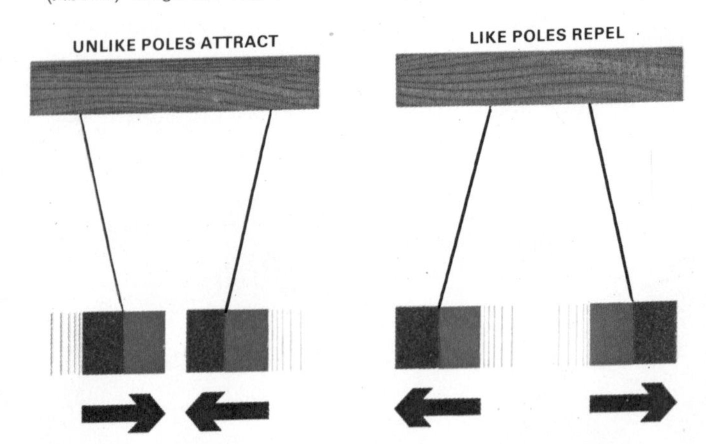

MAGNETISM

As stated earlier, electricity and magnetism go hand in hand and it is necessary to understand one to understand the other. With the finding of lodestone, interest in magnetics arose but with the discovery of iron, and the properties of attraction and repulsion that could be induced in it by rubbing with a lodestone, the science of magnetics was born.

The first property of note about a magnet is that, when freely suspended, it always points in the same direction. As this direction is geographically defined as North–South, the end of the magnet pointing north was christened the north seeking pole or a north magnetic pole, and similarly for the end pointing south. These magnetic poles also attract small bits of unmagnetized iron.

When similar poles on different magnets are brought together they repel each other; opposite poles attract each other. This is the second important property of magnetism.

If a sphere is manufactured from a magnetic material, it is still found to have a north and a south pole. This led to the important concept that the Earth itself is a large magnet having two magnetic poles, one at each end of the axis of rotation. The magnetic pole in the north (commonly called the north magnetic pole) is of south magnetic polarity (by definition) since it attracts the north seeking or *north* pole of a compass needle and *vice versa.* In actual fact, modern measurements have shown that the Earth's magnetic poles do not coincide with the geographical ones about which the Earth rotates. This leads to the navigational term *declination* which is the amount a compass points away from true north at a given point on the Earth's surface. The Earth's magnetic field, as it would be shown by compasses on the Earth's surface, is shown.

The magnetic poles are slowly moving with respect to the geographical ones. At London in 1894 the declination was 17° 0′ West and at present is 6° 47′ West. The mechanism causing the Earth's magnetic field and its movement is unknown. This field is, however, seriously disturbed by magnetic storms on the sun and by showers of charged particles produced by solar explosions.

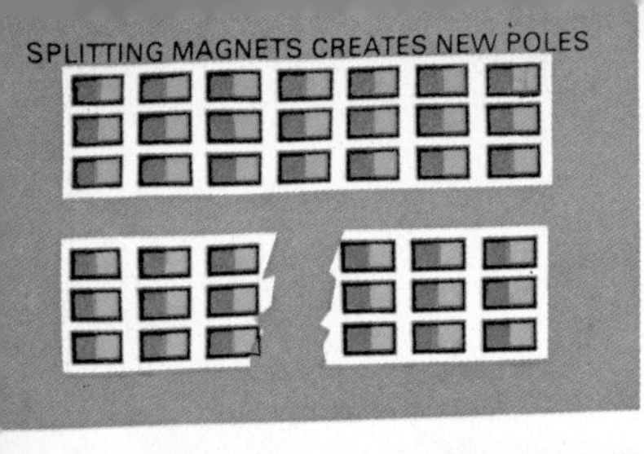

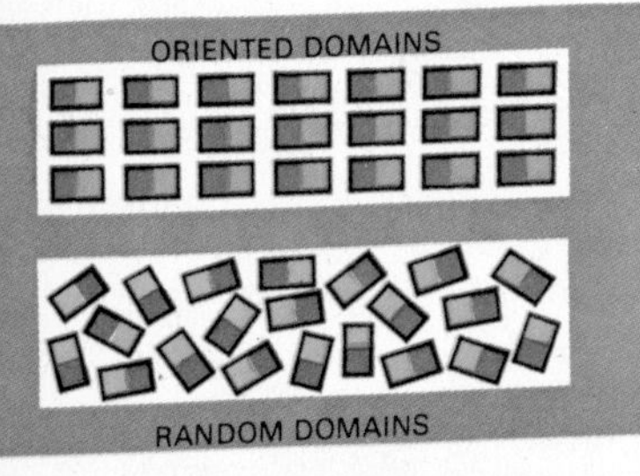

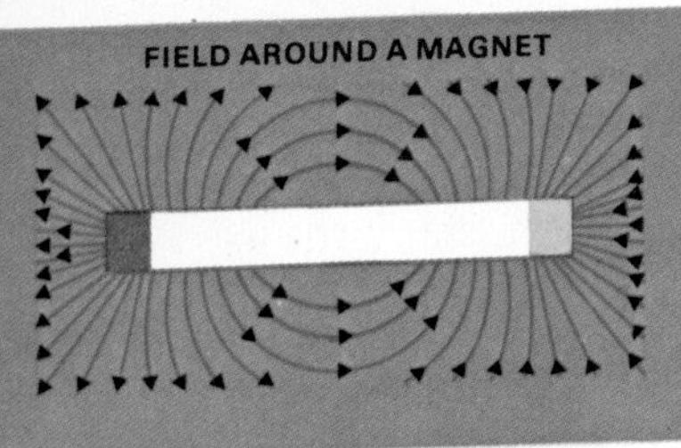

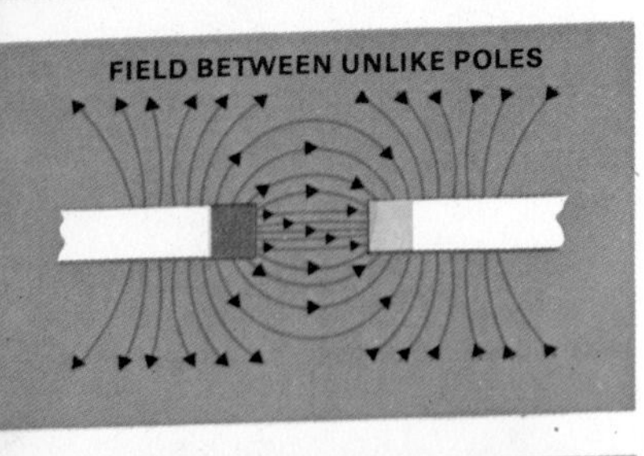

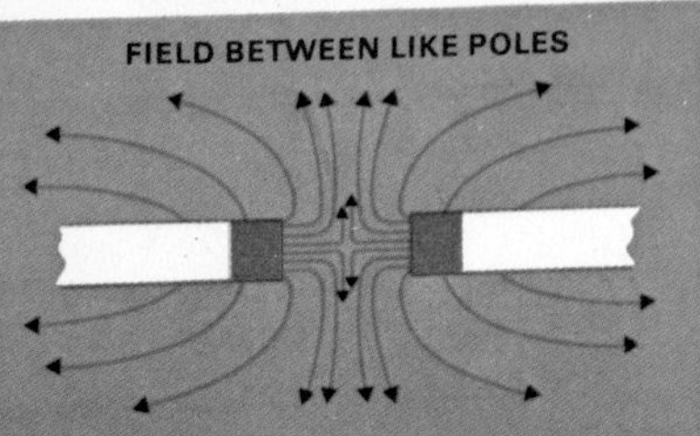

Not all materials can be magnetized. Iron and steel, ferromagnets, were once thought to be the only ones. Later, nickel, cobalt and certain alloys of copper, tin and manganese were found to have marked magnetic properties. Alnico, one of the most powerful magnetic materials, contains 8% aluminium, 14% nickel, 24% cobalt, 3% copper and 51% iron. It accounts for 50% of all magnetic applications at the present time.

Magnetism can be induced in materials in four different ways; (a) from the action of another magnet, by touching or stroking; (b) by hammering a piece of magnetic material held in the direction of the Earth's magnetic field; (c) by heating a piece of magnetic material and letting it cool in a magnetic field; and (d) by passing an electric current through a coil surrounding a piece of magnetic material. Of these, (a) and (d) are the most effective, (d) being the main production method of manufacturing magnets. Conversely, magnetism can be removed by (1) putting it in an opposite magnetic field, (2) banging it, (3) heating it, or (4) passing an opposing current through a coil surrounding the magnet.

Breaking a magnet does not isolate the north and south poles, since new poles of opposite signs appear at the broken ends. This will occur indefinitely, and thus led to the idea that magnetism is a function of the ultimate particles of the material. It is postulated that these particles or *domains* are randomly directed in non-magnetized materials but are aligned in the same direction in a magnet.

A material can only be magnetized until *saturation*, when no matter how high a magnetic field it is subjected to, it will not become any more magnetic. This is due to all the domains being aligned.

Magnetic forces will penetrate non-magnetic materials, so that the magnetic fields set up under different conditions can be observed by placing a sheet of paper over the magnets and sprinkling it with iron dust. The dust becomes magnetized and is light enough to be attracted by the magnetic field.

Because a bar magnet is actually a large number of magnets in a line, the outer surface is bathed in its own magnetic field, which causes it to demagnetize itself. This can be prevented by joining the ends with a soft iron bar (*keeper*) which keeps the magnetic field inside it by providing an easier path than the surrounding air does. Demagnetization can also be prevented by using a horseshoe magnet, so that the field lines no longer cut the majority of the length.

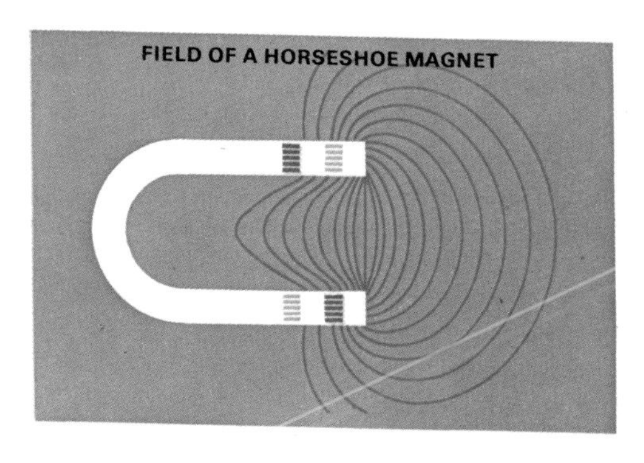

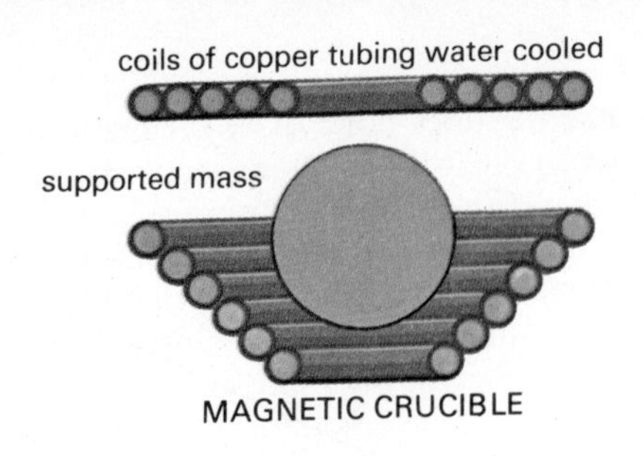

MAGNETIC CRUCIBLE

The basic use of pure magnetism is, as it always has been, for direction finding, since after all a freely suspended magnet always points to the magnetic poles. Compasses range from small hand instruments to the large, permanently mounted types used in ships. The latter types, which have remained virtually unchanged since their introduction in 1900, are about the limit of possible development of the magnetic compass, greater accuracy only being obtained by changing to radio or gyroscopic types. The directions are indicated on a circular card (graduated clockwise in degrees, from 0 at north to 359, with the points N, S, E and W being indicated) which is mounted on a float, pivoted on jewelled bearings and carrying the magnets. Bar magnets are used, usually two mounted side by side, since this double system reduces errors introduced by the ship's motion. The case has a glass top and is filled with a water and alcohol mixture that stops vibration from the ship reaching the float. Lead weights are set in the bottom of the case and the whole assembly is mounted in a double gimbal system so that it remains horizontal, no matter how violent the ship's motion might be.

Electromagnetism is the most important aspect of magnetism in the modern world. It was early noted that a current-carrying wire had a magnetic field surrounding it which could be increased by winding the wire into a coil, and further increased by inserting a soft iron core inside the coil to concentrate the field lines and increase their effectiveness. The magnetism is directly dependent on the coil current, and thus control of magnetic strengths is possible up to saturation point, when an increase in current does not increase the magnetic strength. The magnetism is also dependent on the number of turns in the coil, so that by increasing or decreasing the length of the coil the magnetic strength can be altered. These two facts lead to the concept of the magnetic strength being proportional to the 'ampere-turns', that is, the product of the current and the number of turns in the coil. Thus, for the same magnetic effect, a high current requires few turns and a small current needs a large number.

Electromagnetism has many uses, one of the most important being the lifting of large lumps of iron and steel. A cross-section of the 'lifting head' of an electromagnetic crane is shown. This is very useful with scrap metal, as a number of unconnected pieces can be lifted at one time.

Another use is in relays, which are devices for performing a mechanical action on receipt of an electrical signal. This action normally controls another electrical circuit, for example in a telephone exchange, where hundreds of thousands of relays are used to route calls. A relay is basically a coil formed around an iron core that attracts a pivoted iron plate carrying, or activating, the contacts. A similar device, known as a solenoid, has a hollow coil that attracts an iron plunger. This is used for initiating mechanical action, such as in a remote controlled door catch. The name for such a device is an *actuator*.

(*Below left*) Relay. (*Below right*) Electromagnetic crane head

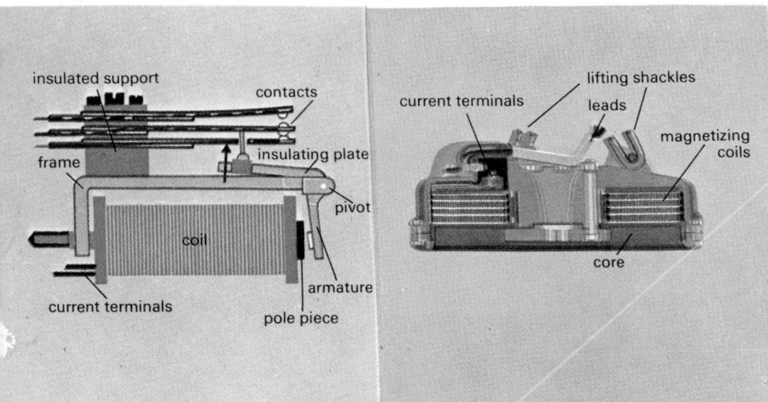

An interesting phenomenon is electromagnetic levitation. Conducting materials can be supported by the interaction of the induced magnetic field in the object and that of the energizing coil, since induced magnetic fields always act to oppose the inducing field – a fundamental law of electromagnetism. This is mainly a laboratory phenomenon but is gradually being applied to industry, where, for example, it is possible for ultra-pure metals to be melted with-

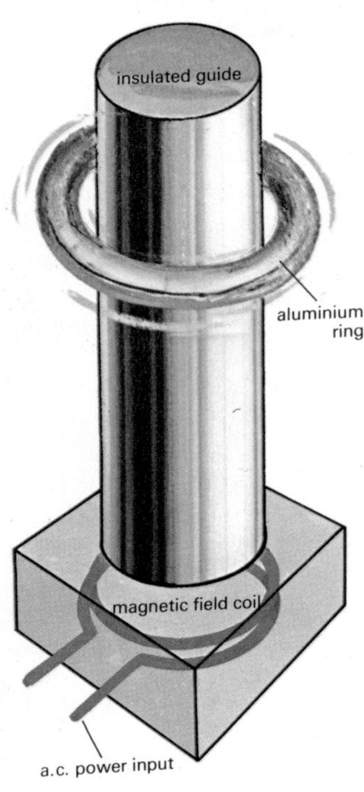

(*Above*) Levitation. (*Right*) Magnetostriction mixing system

(*Below*) Plasma containment

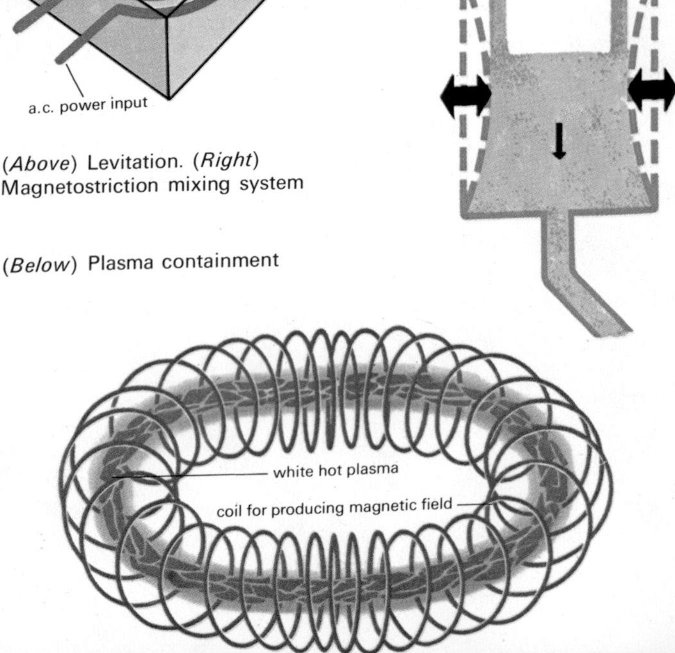

out becoming contaminated, by using magnetic crucibles. These are just a current carrying coil, formed so that the magnetic field set up suspends the metal without it touching the container. Non-metallic impurities fall out as the metal becomes molten.

An interesting and useful, and sometimes troublesome manifestation of electromagnetism is magnetostriction. When the current in the coil surrounding an iron bar is increased slowly, the bar first elongates, then returns to its normal length, and finally decreases in length as saturation is reached. Thus if an alternating current (page 39) is applied, the bar vibrates, the pattern depending on the dimensions of the bar. This effect can be used for mixing light powdery substances, to vibrate containers to remove the contents, and to generate ultrasonic waves (page 127). Very large vibrators based on this principle are used for testing spacecraft instrumentation and systems for reliability. In large transformers (page 62) it is a problem since extremely large mechanical forces can be set up, which may buckle and fracture the assembly.

For studying the basic properties of plasmas, which are possible future power sources, electromagnetic fields are essential. Plasmas are incandescently hot conducting gases, usually hydrogen, in which the electrons have been removed from the nuclei so that both exist separately. Plasmas are too hot (up to 100 million °C) to be contained by any known materials and so they are levitated and confined in powerful magnetic fields.

Magnetometers are another application of electromagnetism. Two coils are wound concentrically around a magnetic core, one being energized by a high frequency alternating current signal, while the other is connected to a sensitive measuring instrument. Any magnetic material brought near the coils will affect the magnetic field slightly and this is indicated by the sensitive instrument. This is the operating principle of mine and metal detectors. In geological surveys, these instruments are suspended from aircraft and used to detect and delineate iron-ore deposits. Such a task would take weeks or months if normal ground methods were employed.

Structure of matter

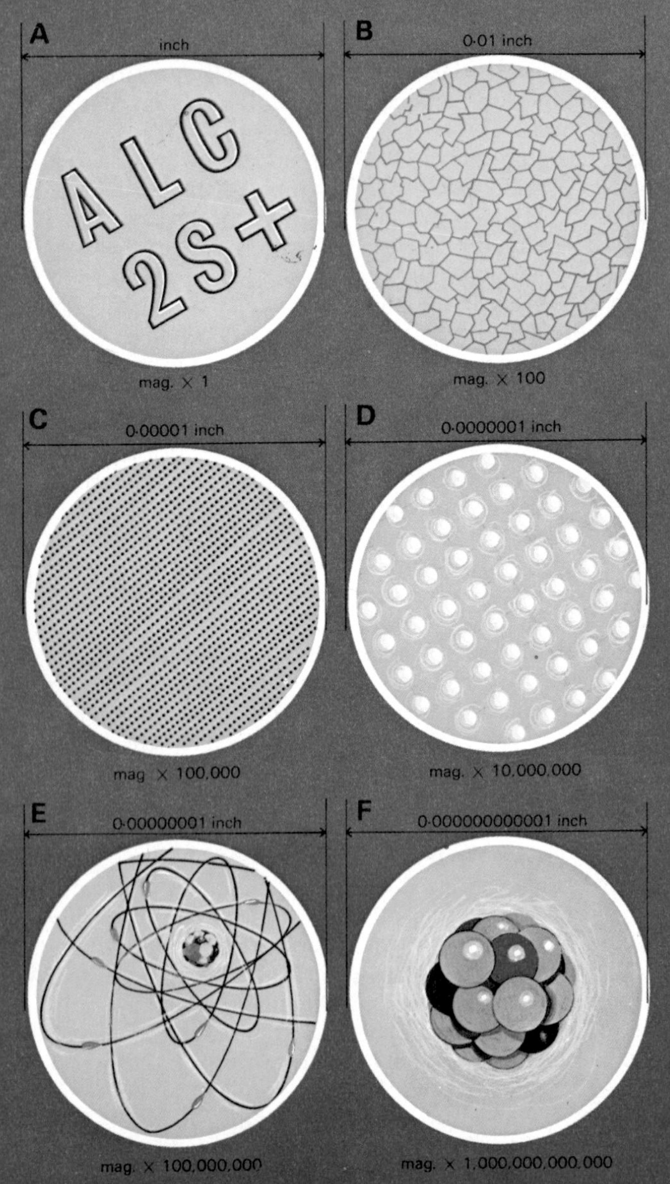

ELECTRICITY

To understand what electricity is, it is necessary to have a knowledge of the structure of matter. Matter is any substance that has weight (mass) and occupies space, for example, air, water or one's body. Three states of matter are possible – solid, liquid and gas. These states are found in nature as elements or compounds which in turn consist of molecules, atoms and sub-atomic particles.

Elements are the basic building blocks of matter and are substances such as iron, carbon or oxygen, which cannot be divided up into other substances by *ordinary chemical means.* To date 105 elements have been proved to exist and it is believed that there are others yet to be discovered. Compounds are chemical combinations of elements, for example water, which is composed of oxygen and hydrogen.

Atoms are the smallest parts of an element that can partake in a chemical combination. Atoms of a particular element have the same mass – atoms of different elements have different masses. Similarly to the way words are formed from the letters of the alphabet, so thousands of compounds can be formed from chemical combinations of the 105 elements. An atom is composed of smaller particles, the principal ones being electrons, protons and neutrons, and can be thought of as being a miniature solar system. The nucleus (composed of protons and neutrons) is at the centre and electrons circle around it like planets around the sun.

The relationship between matter and its constituents is set out in the drawing, which shows what aluminium would look like when viewed through a microscope at increasing powers of magnification. The first two cases are actual, the remainder being hypothetical but based on current scientific principles. While A is what the unaided eye would see, B is the crystalline form of the material, revealed by magnification. In C the presence of individual atoms is revealed and in D these are clearer but fuzzy, due to the outer orbiting electrons merging. A single atom with 13 orbiting electrons is shown in E and finally a nucleus with 13 protons and 14 neutrons is shown in F, looking somewhat like a bunch of grapes.

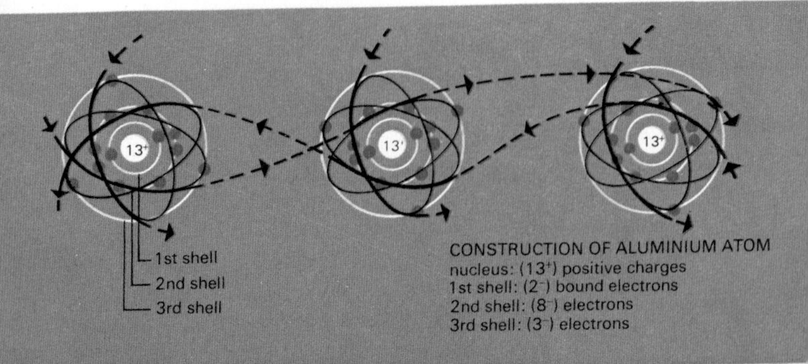

Free electron motion. The dotted line represents the free electron path.

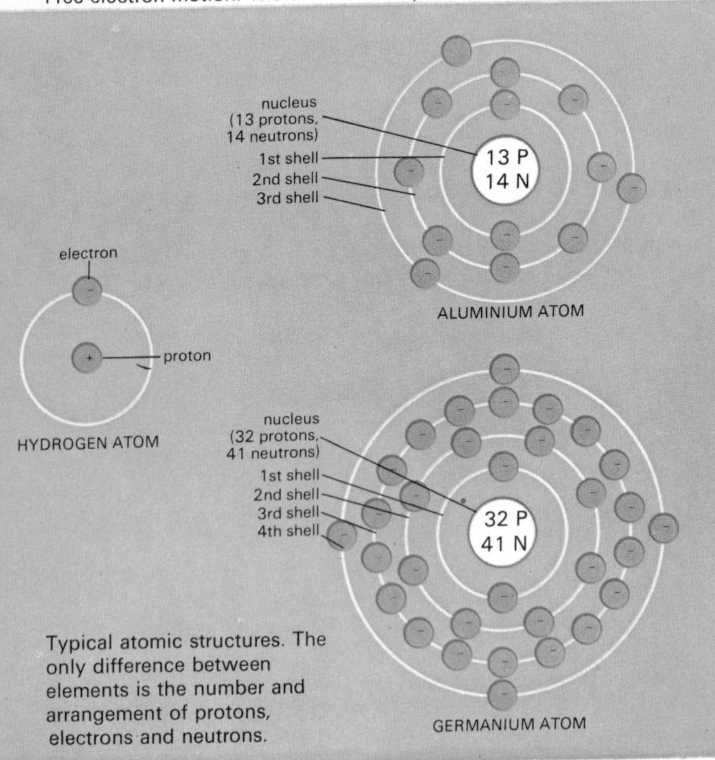

Typical atomic structures. The only difference between elements is the number and arrangement of protons, electrons and neutrons.

The electron has been arbitrarily defined as having a negative polarity and the proton a positive one. An electron is referred to as the *elemental charge* since it is the smallest particle yet discovered to have an electrical charge. Protons, which are 1836 times as heavy as electrons, have a positive charge of equal magnitude. Neutrons are neutral particles, having no electric charge but the same weight as protons. Atoms are normally composed of equal numbers of protons and electrons and are thus electrically neutral. Neutrons add 'weight' to an atom, and there can be more or less than the number of protons in the nucleus. They are not important electrically.

Sometimes an atom gains or loses an electron to become *ionized*. If it gains an electron, it is known as a *negative ion* since it has a negative charge and similarly, losing an electron makes it a *positive ion*.

In certain atoms the electrons in the outer orbits are not tightly bound to the nucleus and can move from one atom to another. Materials containing these *free electrons* are known as *conductors*. Normally this electron interchange is random but, if an electric field is set up by connecting the ends of the conductor across a battery, the electrons start moving towards the positive terminal, the vacant spaces being replenished by electrons from the negative terminal. This flow of electrons is an electric current.

A good conductor requires a large number of free electrons. Metallic elements, such as silver, copper and aluminium are the best conductors. Copper is used most but aluminium, although only 60% as good a conductor, is used in large quantities, since it is cheaper. Silver is the best material, but because of its expense is not in common use. Gold is as good as copper and does not tarnish, but is only used where reliability is much more important than cost, for example in spacecraft.

Materials, such as glass, porcelain, rubber, cotton and plastic, have no free electrons and no inter-atom electron movement occurs. These are known as *insulators*.

Between these two types of materials are those known as *semiconductors*. Their use, for example in transistors, is described in the companion volume, *Electronics*.

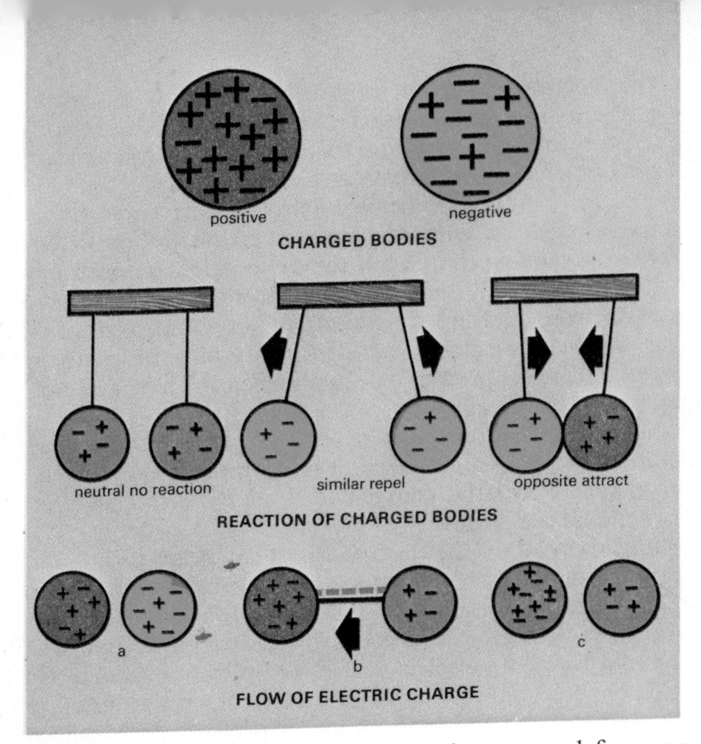

In small quantities, electrons can be removed from or added to insulating materials which become positively or negatively charged, respectively. For example, if a glass rod is rubbed with a silk cloth, the localized heat of friction causes some of the orbiting electrons in the outer atoms of the glass rod to become *free*. These are transferred to the silk cloth leaving a positive charge on the glass rod.

Like magnets, charged bodies attract 'opposites' and repel 'similars'. However, unlike magnets, if oppositely charged bodies are joined together, charge flows from one to the other and the bodies are neutralized. As the only free particles are electrons, this charge flow is an electric current and can be seen by the deflection of a sensitive galvanometer (page 70) when the bodies are first joined. Early experimenters noted that the greater the charge, the larger the deflection of the compass needle, and thus the larger the current flowing.

To cause an electric current to flow between two points, a difference in charge is needed. This charge difference is known as the *potential difference* between the two points since it will cause a current flow if a conducting path is available. *Voltage* is the term commonly used to describe this difference, and it is a measure of the charge available to cause current flow. The unit of voltage is the *volt* (named after Volta). The *current* is then the rate at which the electrons flow in the conductor joining the two points, the flow magnitude depending on the difference in charge. An *ampere* (A) is the unit of current (named after Ampère). This can more easily be visualized by thinking in terms of the illustrated water analogue.

Before the *actual current flow* due to *electron motion* had been discovered (*circa* 1900) the accepted theory was that current flowed from the positive *high potential* to the negative *low potential.* All practical theories and relationships between current and voltage and between electricity and magnetism were based on the original concept and for general electrical engineering purposes this *positive to negative* concept has been retained. The results are the same whichever convention is chosen. Scientists, who are more interested in the fundamental causes rather than the effects of current flow, use the *electron motion* concept, while in the majority of writings on electricity, unless specifically defined as *electron flow*, all currents are assumed to flow from positive to negative and this convention is used in this book.

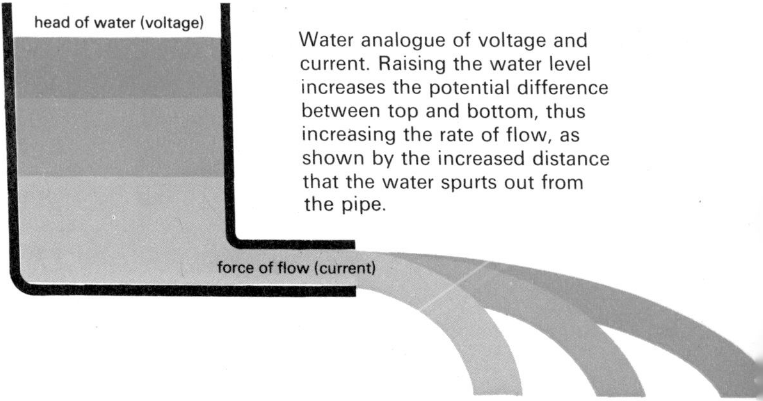

head of water (voltage)

force of flow (current)

Water analogue of voltage and current. Raising the water level increases the potential difference between top and bottom, thus increasing the rate of flow, as shown by the increased distance that the water spurts out from the pipe.

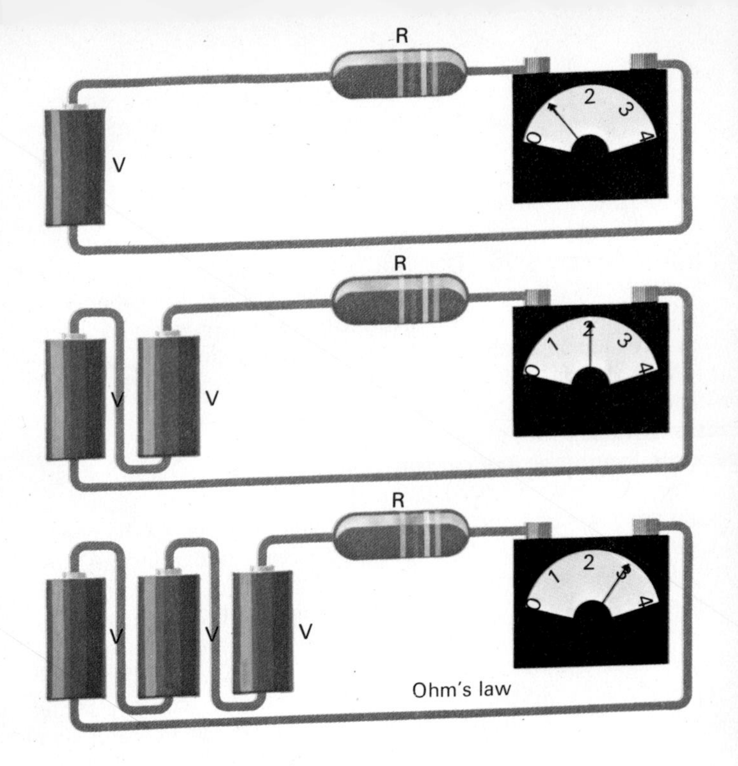

Ohm's law

Georg Ohm, in 1826, discovered that for a fixed conductor size the current increased in equal amounts as additional batteries were added to the circuit, that is, the conductor had a constant resistance to the flow of current which was independent of the applied voltage. Thus the voltage (V) and the current (I) in any circuit are related by a constant known as the *resistance* (R) of the circuit. This empirical fact, (symbolically $I = \frac{V}{R}$) is known as Ohm's law and is the fundamental relationship of electric circuits.

The resistance of a material depends on its size and shape but all materials have a basic characteristic called *resistivity* which depends solely on its composition. Resistivity (ρ) and resistance are related by the formula $R = \rho \frac{L}{A}$ where L is the length and A the cross-sectional area.

Resistance is measured in *ohms* and 1 ohm is defined as the resistance of a conductor passing 1 ampere from a source of 1 volt. The Greek letter Ω is used as an abbreviation. (For example, 10 ohms is written 10Ω.) Resistivity, being a fundamental property of the material, is measured in resistance multiplied by a length, the most usual unit being the ohm–centimetre (Ω–cm).

Some typical resistivity values of various materials are: silver 1·63μΩ–cm, copper 1·77μΩ–cm, aluminium 2·66μΩ–cm, iron 10 μΩ–cm, glass 10^6MΩ–cm and polythene 10^{12} MΩ–cm. This illustrates the difference between conductors and insulators and also why copper and aluminium are the most favoured conductors.

Resistivity, and therefore resistance, is very dependent on temperature; increasing markedly as the temperature rises. Atoms in a material are normally in a state of constant vibration due to the electrons orbiting them. The extent of the vibration increases as the temperature rises. The more strongly the atoms vibrate, the harder it is for the electrons to flow and it requires more energy to keep an equivalent number of electrons flowing. If the temperature is lowered, the vibration decreases, electrons can move more easily and the resistivity decreases.

Because conductors have a very low resistivity the term *conductivity*, which is a measure of the ease with which a current can flow, is used. It is the reciprocal of resistivity and the units used are *mhos* (ohm reversed!). The better the conductor the higher its conductivity.

(*Left*) Measuring *R*, *L* and *A* enables *ρ* to be calculated.
(*Right*) Resistivity increases as the temperature rises.

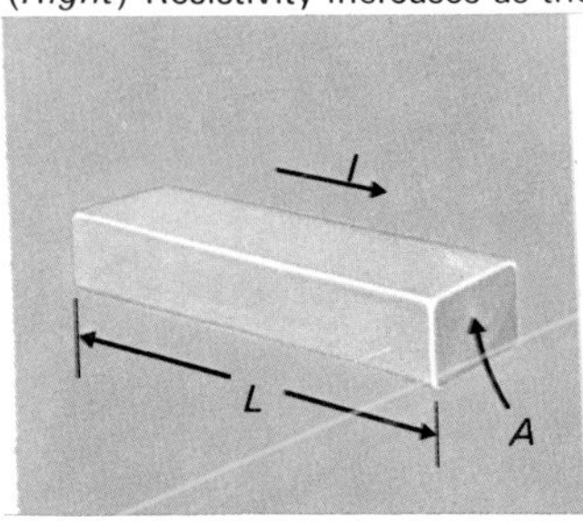

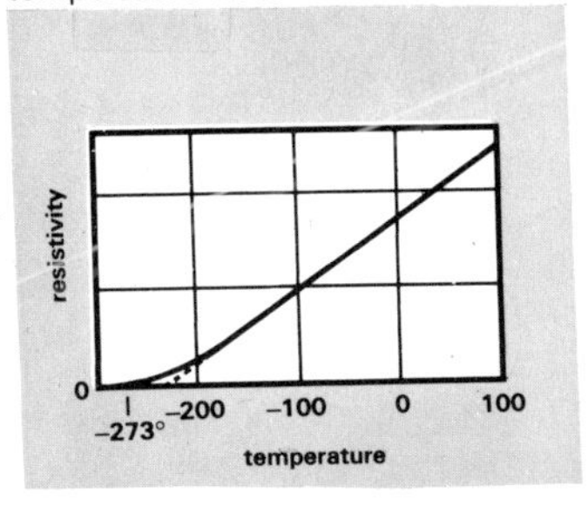

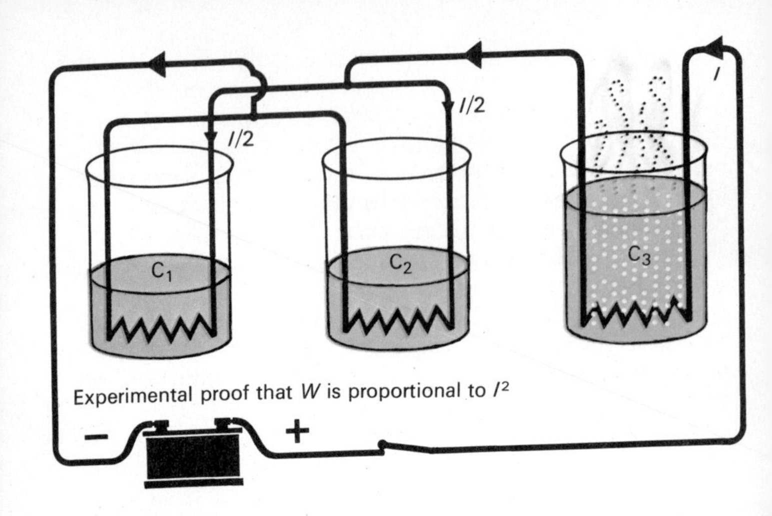

Experimental proof that W is proportional to I^2

The flow of electrons is useful in four ways.

Heating effects

As the free electrons move in a conductor under the influence of the applied electric field, they are repelled by the electrons remaining in orbit and so they do not travel in a straight line. Near collisions result and slow the electron down and it loses energy as heat, so raising the conductor temperature.

The temperature rise depends on the heat generated, which in turn depends on the energy or power being put into the conductor. Power (W) is the product of the current flowing and the voltage applied to the conductor ($W = VI$). By combining this expression with Ohm's law we get $W = I^2R$. In words, power is the product of the resistance and the current squared. Thus, if the current is doubled, the power increases by a factor of four. A simple experiment, using three coils with the same resistance (R) verifies this. The current passes through C_3 and then splits, half passing through C_1 and half through C_2. Now the power generated in C_3 is I^2R but the power generated in C_1 and C_2 is only $(\frac{I}{2})^2R = \frac{1}{4}I^2R$, and thus, although there is twice the amount of water in jar 3 it has four times the heat and it boils long before the other two.

The unit of power is the *watt* (after James Watt) and 1 watt of power is produced when a current of 1 ampere flows through a resistance of 1 ohm. Normally the kilowatt (kW), which is 1000 watts, is used.

Work done by an electric current also depends on the length of time that the power is being dissipated. The unit is the kilowatt–hour (kWhr), which is 1 kilowatt for 1 hour. Thus 1 kW for 2 hours is 2 kWhr. This kilowatt–hour is the unit that electricity meters in the home measure.

Magnetic effects

Magnetic effects of current flow are the basis of instruments used for detecting and measuring electric currents and of the operation of electric motors.

Chemical effects

Electric currents can cause the separation of various chemical compounds into their constituent elements or help combine others and this is covered in later sections.

Charge storage effects

Current flow can remove or deposit *charge* on an object and this is discussed in future sections.

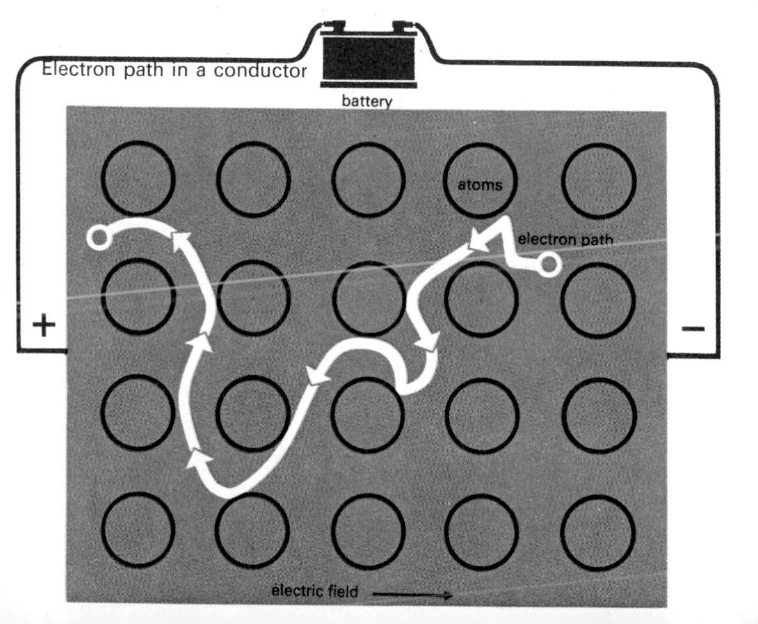

Electron path in a conductor

Production of electric currents

Electric currents can be produced in five ways: by flow of static electricity; chemical reaction; animal electrification; thermoelectricity; and cutting of magnetic lines of force. These methods are described in the following sections.

Flow of static electricity

This is the earliest known form of electricity. When a charged object is brought near an earthed one, a small spark jumps the gap. This spark is a flow of electric current.

Most people are well acquainted with this phenomenon. Who has not at one time or another walked over a thick carpet and then received a shock upon touching a metal object or another person? This is especially apparent if man-made fibres, such as nylon, are involved.

Static electrification is a great danger in industry where a slight spark can cause a bad explosion, for example in an oil refinery. Hospital operating theatres are very prone to this danger, as most gaseous anaesthetics are highly explosive. Static electricity can also be dangerous in spacecraft which use volatile fuels. A number of anti-static materials have been evolved to combat this hazard. They are slightly

conducting so that the charge will leak away instead of gradually building up to a level great enough to cause a spark. A dry atmosphere is dangerous, since dry air is a good insulator, but if the atmosphere is moist charges leak away as moisture is a fairly good conductor.

Static charges can be useful and are one of the best and simplest sources of generating high voltages for scientific purposes. A van de Graaf generator operates by charge being sprayed on to an insulated moving belt and carried up to the top metallic dome. Opposite charge is forced out of the dome and is removed by the belt to earth. This is a continuous process and, as the charge cannot leak away, very large voltages, up to 4 million volts (4 MV) can be built up. Large currents cannot be produced, since only a limited amount of charge can be transported by the belt at any one time. So generators of this type are normally used for generating X-rays and accelerating particles in *atom-smashers* where high voltages, not currents, are important.

van de Graaf generator

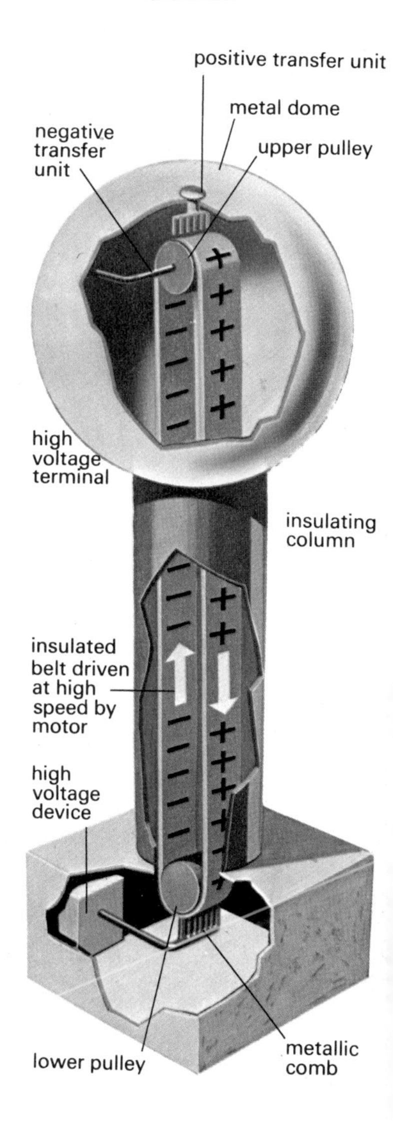

(*Left*) Effects of static electricity

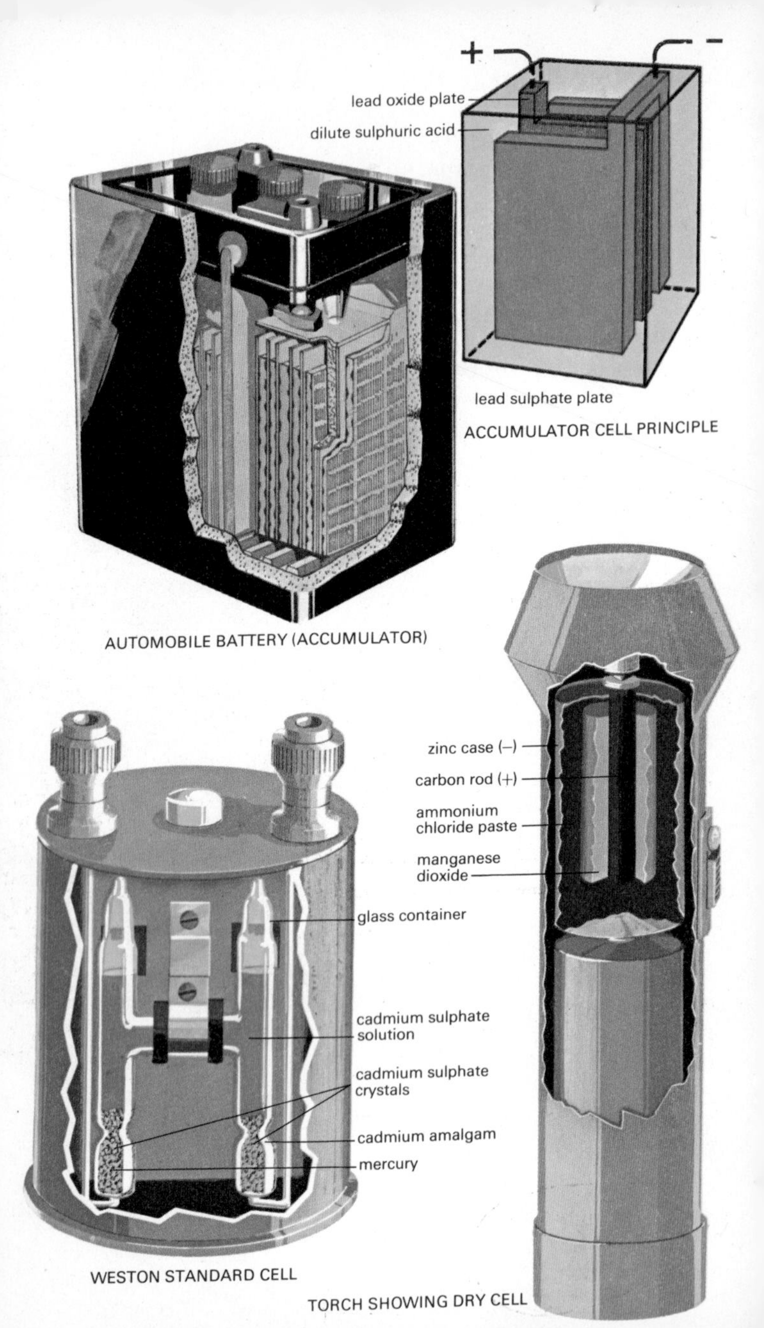

ACCUMULATOR CELL PRINCIPLE

AUTOMOBILE BATTERY (ACCUMULATOR)

WESTON STANDARD CELL

TORCH SHOWING DRY CELL

Chemical reaction

A metal plate in an electrolyte (conducting liquid) reacts chemically, emitting positive ions (page 23) and thus becoming negatively charged. If a plate of different metal is immersed in the same liquid and joined externally to the first a current flows, since the two metals have different ion emission rates and therefore different negative charge values. This interrupts the ionic equilibrium of each metal and charge transfer occurs in the liquid until all the available ions are used up and the 'cell' is run down.

This is the basis of the lead acid storage battery or accumulator, as used in cars. A typical battery consists of several cells, containing plates of lead dioxide interleaved with plates of pure lead in a solution of dilute sulphuric acid. The plates of each material are attached together and, if joined to the plates of the other material, a current flows through the connection. A chemical reaction occurs, both materials gradually changing into lead sulphate. When the ion exchange rates become equal, no charge difference exists and no current flows. The battery is now discharged but this type, a secondary cell, is rechargeable. If a current is passed through the battery in the opposite direction, the plates return to their original condition. They become charged again. Cells of about 2 volts are connected in series to give a higher voltage. Higher currents are obtained by connecting in parallel or using larger plates. Batteries are usually rated by the current (amperes) that they can deliver for a given time (hours). For example, a 40 ampere hour unit can give 40 amperes for 1 hour or alternatively 10 amperes for 4 hours.

An example of a non-rechargeable or primary cell is the torch battery. This dry cell has a wet paste electrolyte that dries out as it gradually discharges. The complete cell is then replaced.

Another primary cell is the *Weston Standard Cell*, the voltage of which can be accurately calculated from the chemical properties of the materials used. This cell was used as an international standard of potential and is still used as a laboratory reference cell. Its use is limited to this, since very little current can be drawn before discharge.

Animal electrification

All living creatures chemically develop electric potentials, most using them only for conducting 'messages' throughout their nervous systems. In humans, for instance, the electrical pulses travel along nerve fibres at 1 to 100 metres per second. These fibres are uni-directional, about one micrometre in diameter, and consist of an insulating membrane surrounding a liquid conducting core. A stimulus effects a temporary change in the membrane. Potassium ions from the slightly negative core pass outwards through it, leaving a local positive charge on the core. The current, induced by the potential (about 0·12 volts) between this *active*, positive section and the negative *resting* core ahead, activates the next section of membrane, thus propagating the pulse. Behind the pulse potassium ions drift back into the core, reaching their former resting concentration in about one millisecond, this time being the *latent period*, during which the nerve cannot respond to another stimulus.

A few fish use electricity offensively in hunting, their prey being stunned by an electric shock. A large electric eel can administer a pulse of 600 volts at a few amperes for 3 milliseconds, strong enough to stun a swimmer. Much of the body is composed of thousands of disc-like cells of muscle material, connected in series in many parallel rows. The potential is generated by the slow diffusion of potassium and sodium ions through the cell walls in opposite directions (each type can only pass in one direction). On discharge, the cell walls become 'open' to flow in the opposite direction and the ions intermingle. The discharge is controlled by the central nervous system and can be released at will. The nose is negative and the tail positive and both must touch the victim for a discharge to occur. The evolutionary history of these electric organs remains an unanswered but extremely interesting question.

Another chemical reaction in living tissue, now being commercially exploited (page 154), causes the emission of 'cold' light. The firefly exhibits this *electro-luminescence*, using it to attract the opposite sex. Chemicals in the body react with oxygen in the air and glow intermittently. A steady glow can be obtained in pure oxygen.

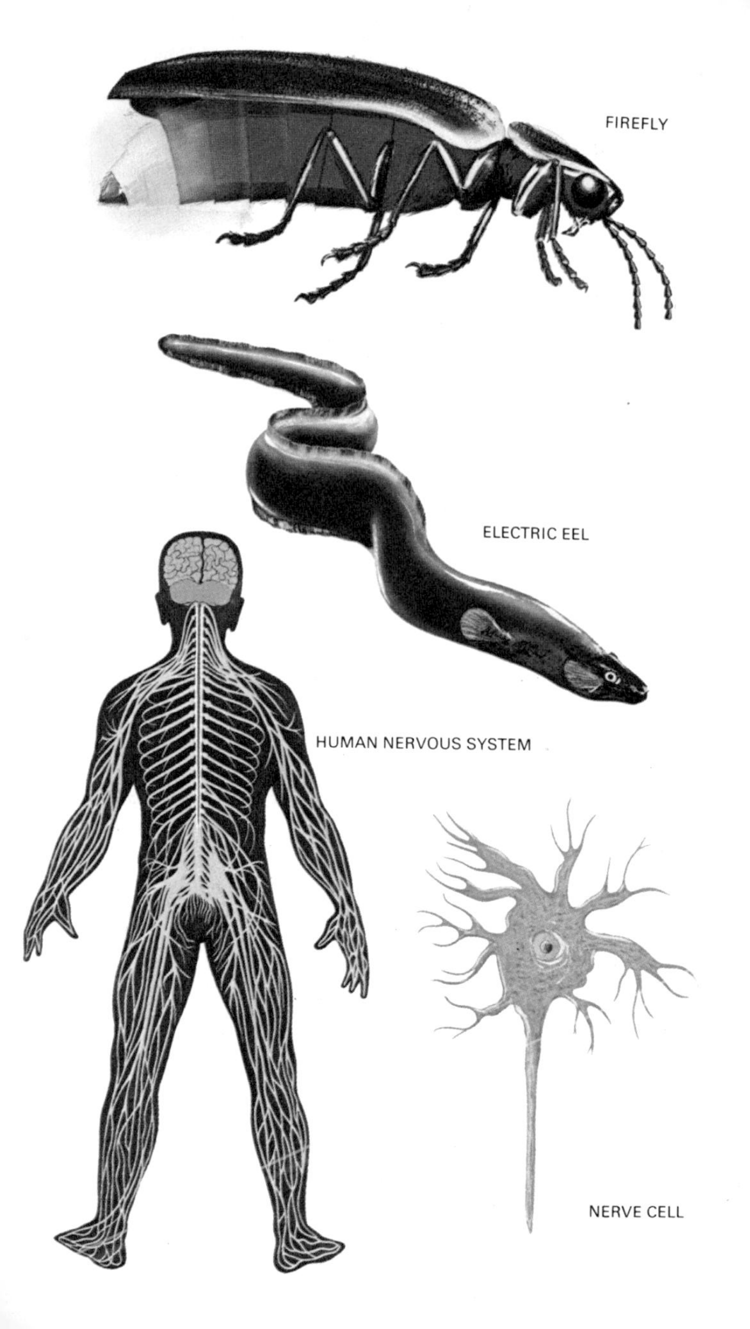
FIREFLY
ELECTRIC EEL
HUMAN NERVOUS SYSTEM
NERVE CELL

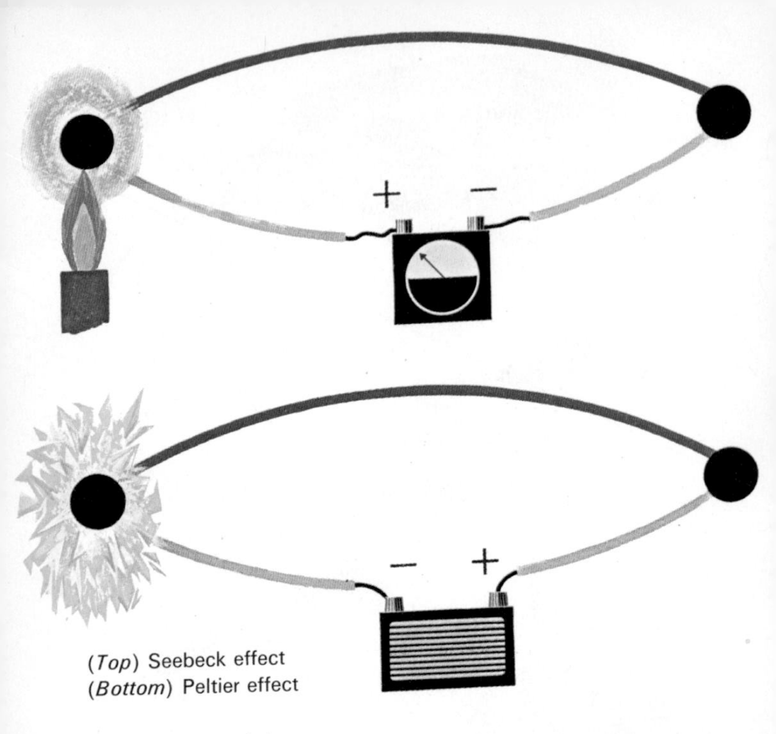

(*Top*) Seebeck effect
(*Bottom*) Peltier effect

Thermoelectricity

In 1826, Seebeck observed that if one junction of a loop, consisting of two dissimilar metals (for example antimony and bismuth), were heated, an electric current flowed. An explanation is beyond the scope of this text but the basic properties of the phenomenon are that (1) the current flow is always the same for any given temperature difference provided that the metals are the same; (2) the current flow and thus the voltage available is proportional to the temperature difference between the junctions; and (3) if a number of junctions are in series, the voltages add.

The second fact is one of the most useful, since it is used to measure temperatures in inaccessible places. A thermocouple is a loop of two metals, such as copper/constantin, set up with one junction on the object to be measured and the other kept at a known reference temperature, usually 0°C with a melting ice bath. The potential difference between the junctions is measured and the object temperature found.

Very small temperature differences can be measured using a thermopile, which is based on the third fact. It consists of a large number of junctions connected electrically in series, but thermally in parallel. This gives a large change in potential for a small change in temperature.

The corollary to this phenomenon is the Peltier effect: when a current is passed through a loop of two strips of dissimilar metals, one junction becomes colder and the other hotter. This effect can be swamped by the normal heating effect of a current in a wire unless a large number of junctions are thermally paralleled. It has found limited use in small refrigeration and air conditioning equipment. A useful application is the providing of a reference junction at 0°C for a thermocouple. Here as the water cools and ice forms, the expansion in volume of the enclosed liquid stops the process by activating the microswitch and, as the liquid contracts on warming, the switch closes and cooling restarts.

Temperature measurement by thermocouple

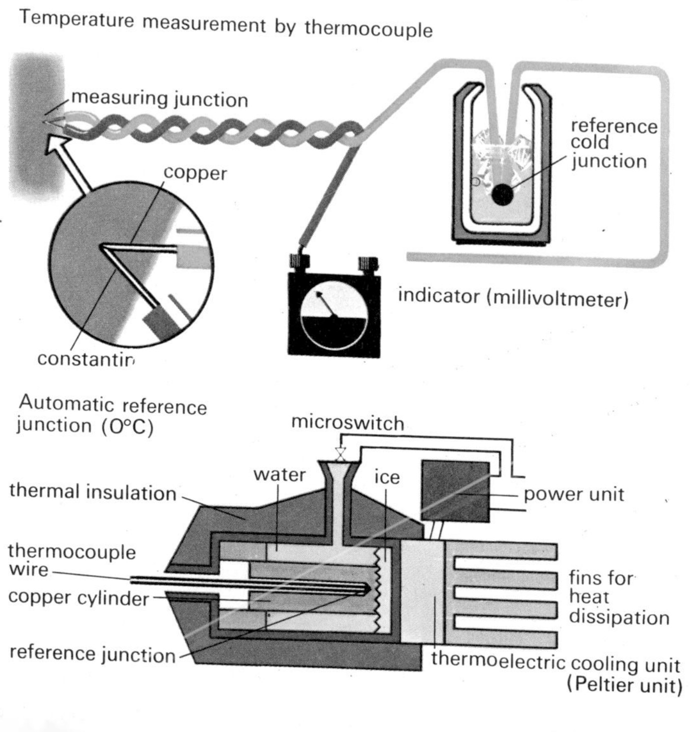

Automatic reference junction (0°C)

Cutting of magnetic lines of force

This is the most important method of producing electricity, over 99·9% of the electricity used in daily life being obtained in this manner. Only in this way can mechanical energy be converted to electrical energy, large and continuous amounts being thus produced without, for example, the need to recharge a battery. Faraday was the first to show that if a conductor is moved across a magnetic field, or *vice versa*, an electric current will flow in the conductor if it is connected by an external circuit. Later Fleming noted a simple relationship, the *right hand rule*, which states that if the thumb, forefinger and second finger of the right hand are held mutually at right angles with the thumb indicating the direction of motion and the forefinger the magnetic field (N to S), then the second finger indicates the direction of current flow.

Machines that produce electricity in this way are known as generators. Rotational motion is generally the easiest to produce, and so generators work on the principle of rotation of a wire coil through a magnetic field, or a magnetic field past a wire coil. A simple generator consists of two magnets having opposite poles adjacent (or a horseshoe type) with a coil of wire, which is free to rotate, mounted between them. The ends of the coils are brought out to metal collector rings, called *slip-rings*, the current being picked off these rings by fixed sliding contacts, called *brushes*, composed of carbon. The coil is rotated by mechanical means.

Fleming's 'right hand rule'

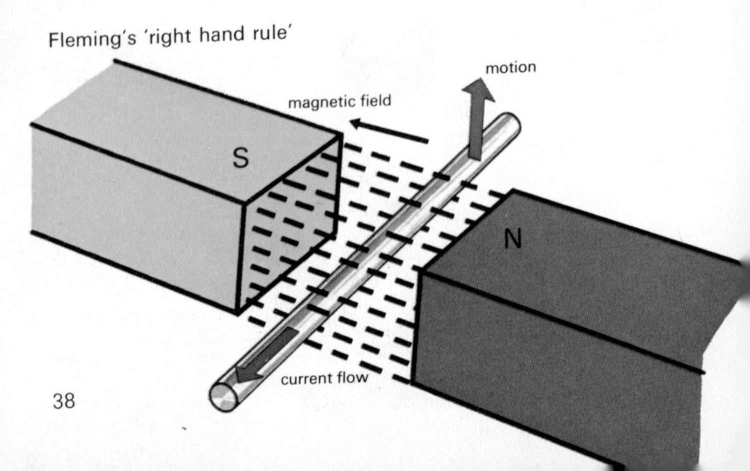

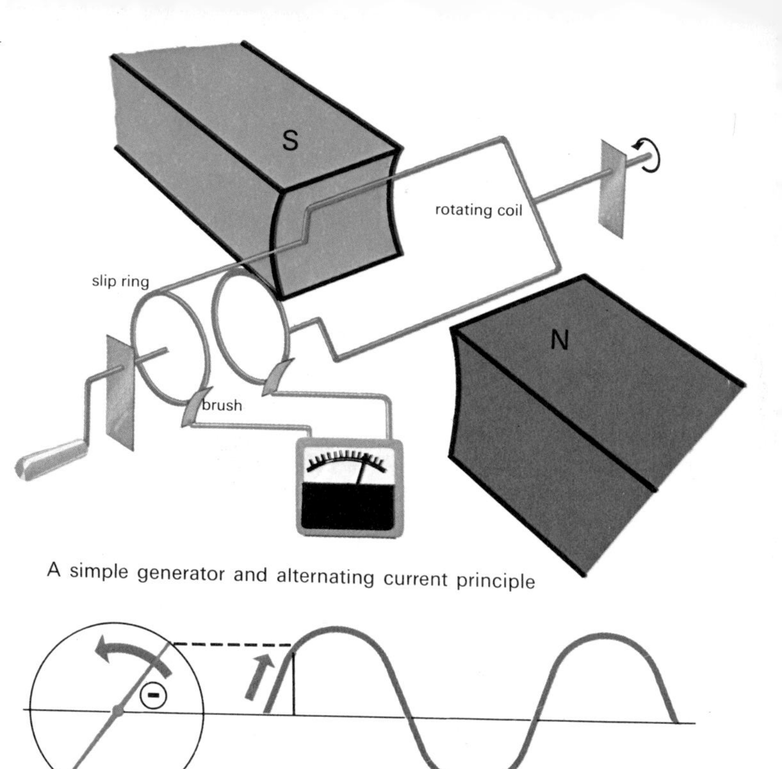

A simple generator and alternating current principle

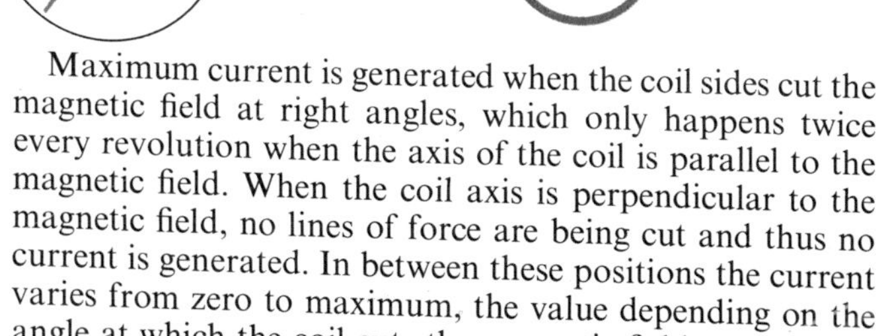

Maximum current is generated when the coil sides cut the magnetic field at right angles, which only happens twice every revolution when the axis of the coil is parallel to the magnetic field. When the coil axis is perpendicular to the magnetic field, no lines of force are being cut and thus no current is generated. In between these positions the current varies from zero to maximum, the value depending on the angle at which the coil cuts the magnetic field.

As the coil rotates the sides cut the magnetic field alternately in opposite directions. Thus the induced current changes direction every half revolution and an *alternating current* (a.c.) is produced. The current increases from zero to maximum in one direction, falls to zero again and then reverses, following an identical pattern in the opposite direction and this is repeated at every revolution of the coil.

A simple generator produces an alternating current (and thus a voltage) that repeats itself every revolution, that is, it changes periodically with time. Each repetition is known as a cycle and the current is defined by the number of cycles occurring every second, called the *frequency*, the unit of which is the *hertz* (after the man who discovered radio waves). One cycle per second equals one hertz (Hz).

The magnitude of the induced current depends on the strength of the magnetic field, the speed of rotation and the angle at which the coil cuts the magnetic field in a vertical direction at a particular moment. The magnitude at any moment (I) is found to be proportional to the vertical distance from the tip of a line (called a *vector*) representing the maximum magnitude or *amplitude* (I_m) rotating at the same speed as the coil. In this example the instantaneous amplitude and the path traced by the current during 150° of rotation are shown. This shape of wave is defined by a mathematical term, the *sine* of the angle between the zero position and the position of the coil (θ), which relates the vertical distance at any position to the vector length. Thus $I = I_m \sin\,\theta$. Because it is described by this mathematical term, this shape of alternating current is said to be sinusoidal and is known as a sine wave. An alternating current is not necessarily sinusoidal; it can be any shape as long as it repeats itself periodically.

The maximum value is not a true measure of the amount

Development of a sine wave

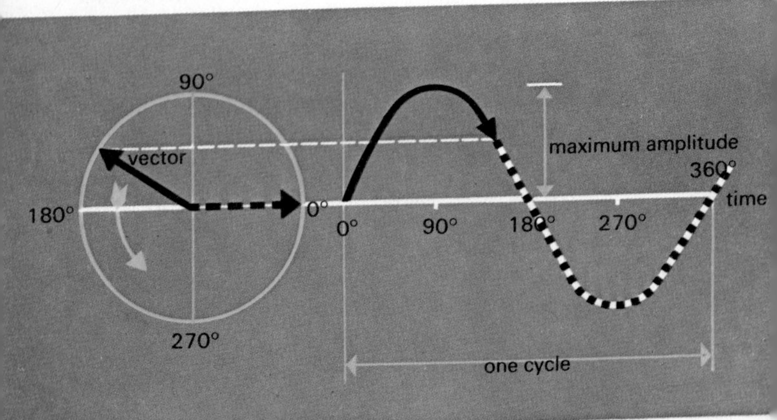

of current flowing, since it only reaches this value twice a cycle, and also two different shaped waves may have the same amplitude, and so the average value should be used. For a sine wave this turns out to be zero! However, reversing the current in a conductor does not cause cooling; it still causes heating since current still flows. Now the power dissipated in a circuit is proportional to the square of the current, and as the square of a negative quantity is positive, the average power over a cycle is positive. The *effective* current over a cycle is thus the square root of the average of the sum of the squares of the instantaneous values of the current and this is found to be 0·707 of the maximum; that is, $I_{eff} = 0{\cdot}707\ I_m$. This is the value most commonly used to measure alternating currents and voltages and so, unless otherwise stated, all magnitudes of alternating current are the effective values.

The *phase* of an alternating wave is the angular (ϕ) or time difference between it and another wave of similar frequency. As will be shown later, the voltage and current in circuits are not always *in phase* (having both their maximums and zeros occurring at the same instance in time) but have a *phase shift*, the amount of which depends on the components used. Thus one wave leads or lags the other by a fixed amount as the two vectors rotate at a constant speed. In phase waves have a zero phase shift and the two vectors are superimposed.

Phase shift

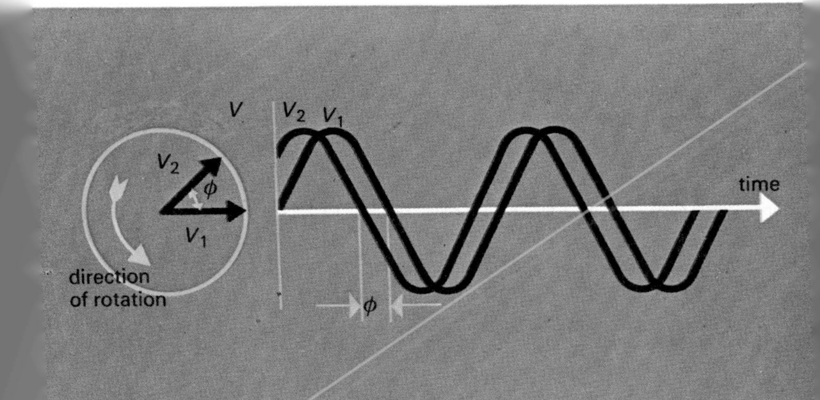

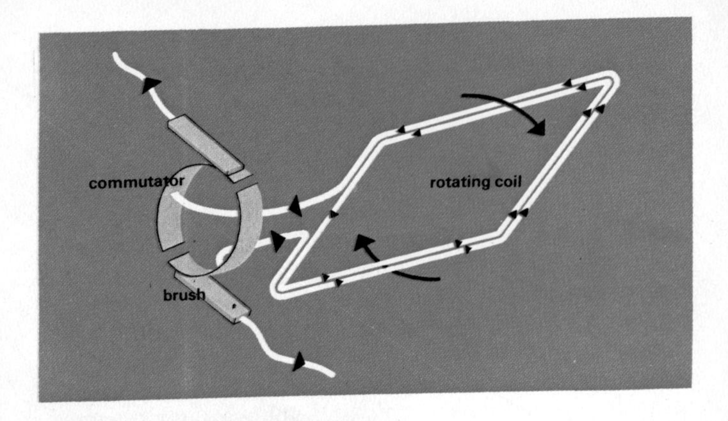

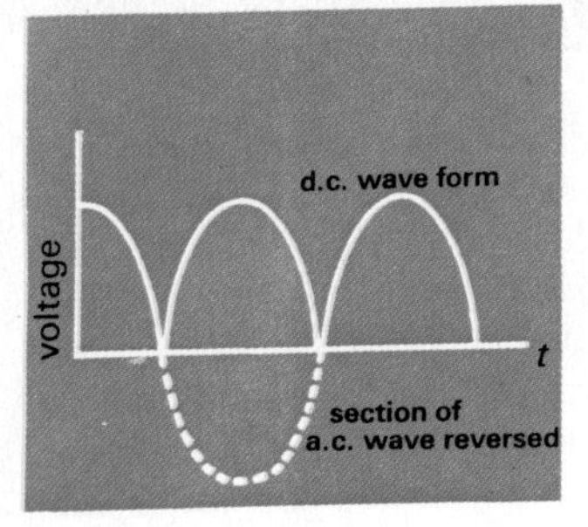

(*Above*) Two section commutator generator. (*Left*) Pulsating direct current from two section commutator

Sometimes a steady (non-alternating) current is needed, for example for electroplating. This always flows in the same direction and is constant in amplitude, being known as *direct current* (d.c.). The first four methods of producing electricity (page 30) give this type of output. A simple mechanical modification to the basic a.c. generator allows direct current to be continuously obtained from mechanical energy.

The slip rings are replaced by a *commutator* which is one ring split in half (usually separated by mica insulation), each half being connected to one end of the coil. In this way the connections to the external circuit are reversed as the current in the coil reverses, thus causing the current in the external circuit to flow always in the same direction.

However, the current is pulsating, changing from zero to

maximum twice per revolution, not steady like that from a battery. If two coils, at right angles to each other, are used with a four section commutator a smoother output results and this can be improved by adding more coils and dividing the commutator into even more segments. Actual machines have up to thirty coils, giving a virtually constant output.

By definition, equal direct and alternating currents generate the same amount of heat in a conductor. Thus $I_{dc} = I_{eff}$.

In practice, electromagnetism from coils wrapped around soft steel formers, rather than permanent magnets, produces the magnetic fields. The currents *exciting* these coils are normally taken from the generator output. This is known as *self-excitation*, a shunt connection being employed (page 111). The soft steel used for the magnetic field poles always retains a slight magnetism when the current is removed. This *residual magnetization* is enough to generate a small current, which increases the magnetic field, in turn causing a larger current and so on until the generator is giving maximum output.

Control of the voltage output is achieved by regulating the field coil current with a variable series resistance. The higher the current the greater is the output.

The brushes are set so that they change from one commutator segment to the next when the coil is in the neutral position, that is, when no current is being induced in it from the field, since otherwise a spark would be caused as the brush left the segment. In practice there is always a slight current present, due to induction from other coils and stray fields and so some sparking results and reduces the life of the brushes and eventually the commutator.

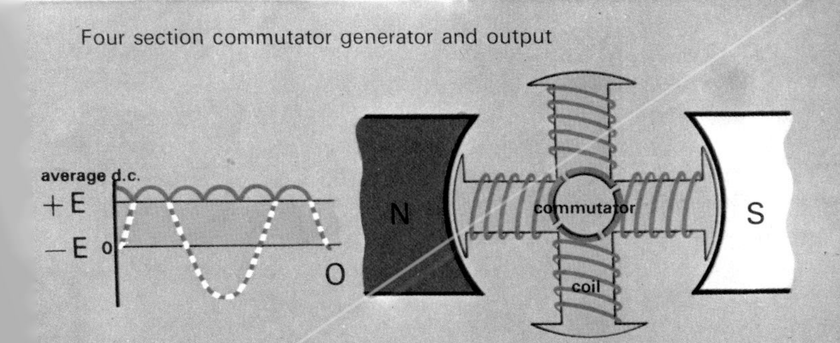

Four section commutator generator and output

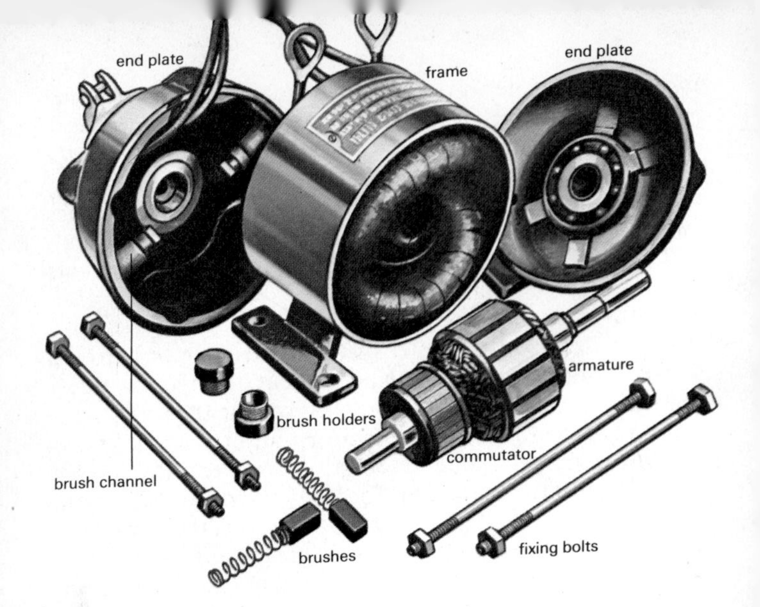

Small d.c. generator

In practical machines the coils (windings) are distributed around the periphery of the *rotor* (the moving part) and the *stator* (the fixed frame). Each coil consists of many turns of wire to give higher output voltages: each turn has the same induced voltage so that the total equals the number of turns in the coil multiplied by the volts per turn. Several coils are connected in parallel to give higher currents.

All d.c. machines have the magnetic field on the stator, since the use of a rotating commutator is required to produce direct currents. This limits the currents and voltages available from the machine, since the small gap between commutator segments limits the voltage to several hundred volts and high currents cause severe arcing damage to the commutator as the brushes move from segment to segment.

Most a.c. generators (alternators) invert the position of the magnetic field, having it on the rotor and the output windings on the stator. Large currents and voltages can thus be generated, since no moving contacts are necessary.

The output of generators is increased by using more than one pair of magnetic poles. There is, for alternators, a relation between speed (n), frequency (f) in Hz of the output voltage and the number of pairs of poles (p), namely $f=\frac{pn}{60}$.

In all machines the generated voltage is equal to $KpnØN$, where K is a constant that depends on the construction of the machine, N is the number of turns per winding and Ø is the magnetic field strength. A small magnetic field can thus produce large amounts of electricity, if a large number of turns and poles are used, and the mechanical input can give high speeds.

Most large d.c. generators are self-excited with provision for controlling the current incorporated in the field circuit. Alternators have a separate source of excitation, usually a small self-excited d.c. generator, operating off the same shaft and connected via a control circuit to the rotor field coils.

Large a.c. generators. (*Above*) 150 MW high speed turbine rotor. (*Right*) 50 MW stator

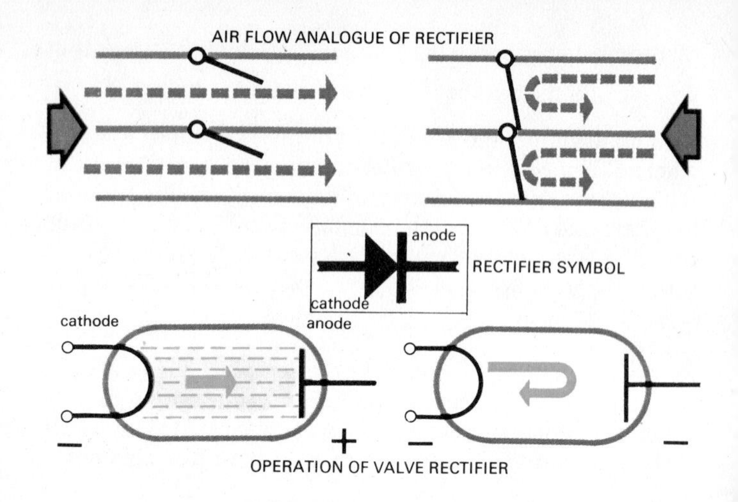

Rectifiers

Although alternating current is much easier to produce and transmit in large quantities than direct current, numerous industrial applications require direct current.

Initially, rotary converters, mains fed a.c. motors driving d.c. generators that fed the load, were used to convert a.c. to d.c., but these were large and inefficient, with limited power conversion potential. The more efficient rectifier permits current to flow through it in one direction only. In the analogue, the air flow is the current and the shutter the rectifier. It can be seen that the air can only pass in one direction. Early rectifiers were glass bulbs containing two electrodes (diodes) separated by the vacuum gap. When one electrode (cathode) is heated, it gives off electrons which, if the other electrode (anode) is positive, that is during one half of the a.c. cycle, move across the gap, causing a current flow. If the anode is negative, during the other half of the a.c. cycle, the electrons are repelled. Thus if a full cycle of a.c. current is fed to a diode only half of it can pass.

It was found that a high vacuum was unnecessary and that low pressure gases could be used if the electrodes were of dissimilar metals. An arc could be formed easily in one direction but not the other. The best materials are carbon for the anode and mercury for the cathode in a glass bulb containing low pressure mercury vapour. This apparatus is known as a *mercury arc rectifier*.

When a material with an excess of electrons is in contact with one having a dearth a junction, passing electrons in one direction only, forms. Early types consisted of a metal cathode (copper) and a semiconductor anode (copper oxide or selenium oxide). Modern *solid state* diodes are junctions of two different semiconducting materials.

Solid state rectifiers are small and cannot handle large powers or voltages. The vacuum diode can be used for high voltages and small powers, while the mercury arc rectifier can handle large powers at moderately high voltages, and is the commonest in use in industry. Solid state rectifiers are being continually improved and are now capturing a larger share of the industrial market.

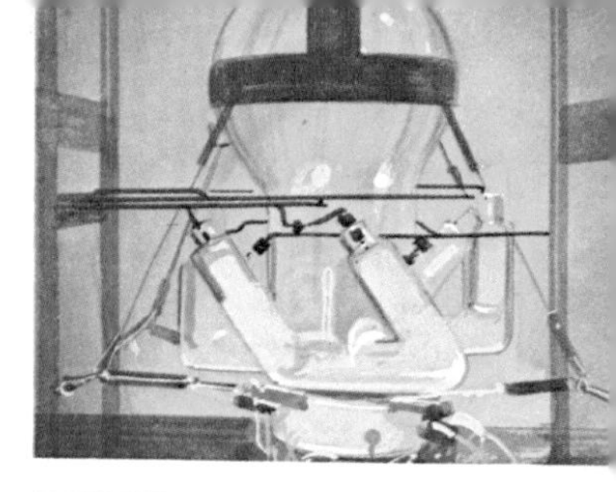

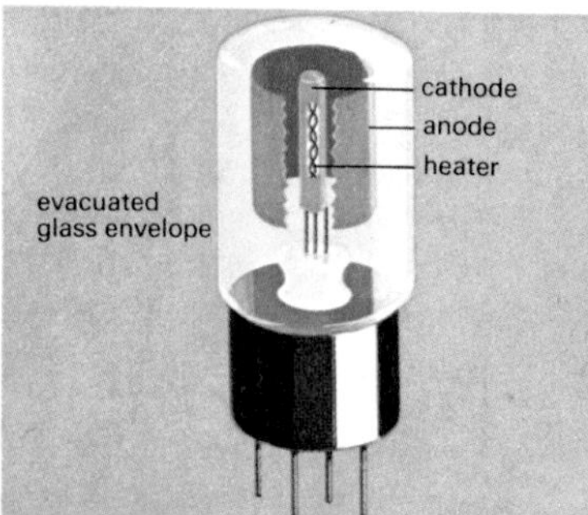

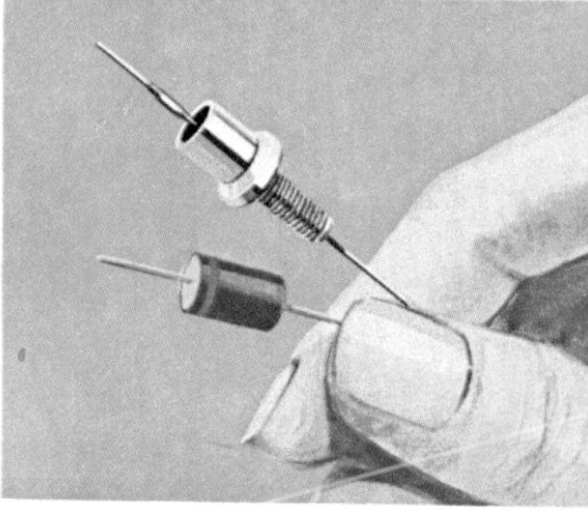

Rectifiers. (*From top*) Mercury arc, valve, semi-conductor and copper oxide

Rectifier circuits

Rectifiers are used in a number of circuit configurations, the simplest being the *half wave* circuit, where only half the a.c. cycle is used. If another rectifier is added and the secondary winding of the transformer doubled, the *full wave* circuit results. This utilizes both half cycles of the supply. The half wave circuit can also be used with a *polyphase* (more than one phase, page 90) supply, up to 24-phase rectification being utilized by suitable transformer connections.

The figures show that the amount of d.c. present increases as the circuit changes from half to full wave to three phase and that the circuit efficiency is increasing. The current produced is not pure d.c. It still has an alternating component called *ripple*. This ripple, characterized by the *ripple factor*, which is the ratio of the a.c. to the d.c. components of the rectified wave, is a measure of the conversion efficiency and the lower it is the better the rectification.

The ripple can be reduced by adding a filter to the circuit. A filter is a combination of series inductances and parallel capacitors. The capacitor charges up to the peak value of the wave and then discharges slowly through the load, while the rectified wave falls rapidly. The inductances also block the a.c. component (page 57). The capacitor is thus storing charges while the rectified wave is present and releasing it

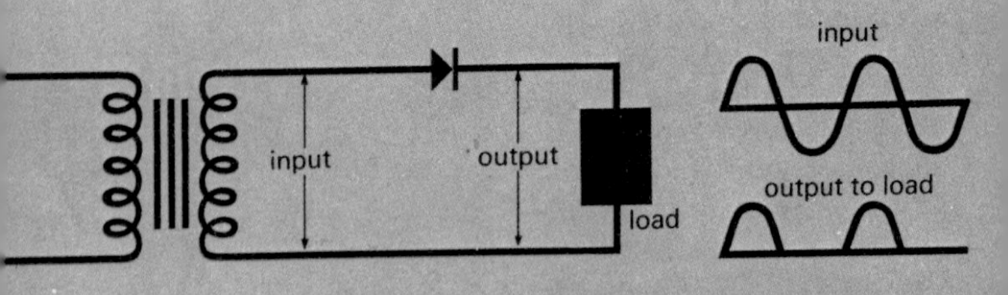

(*Above*) Half wave rectifier. (*Below*) Full wave rectifier

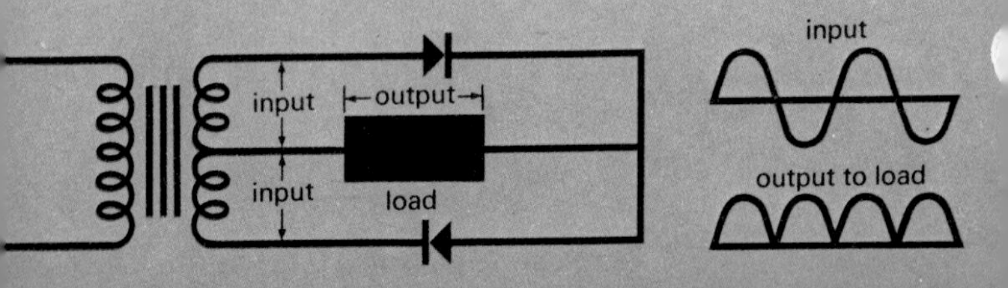

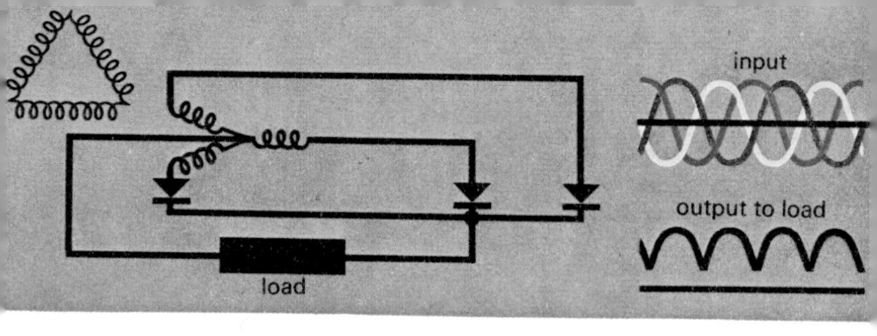

(*Above*) Polyphase rectifier. (*Below*) Filter circuit

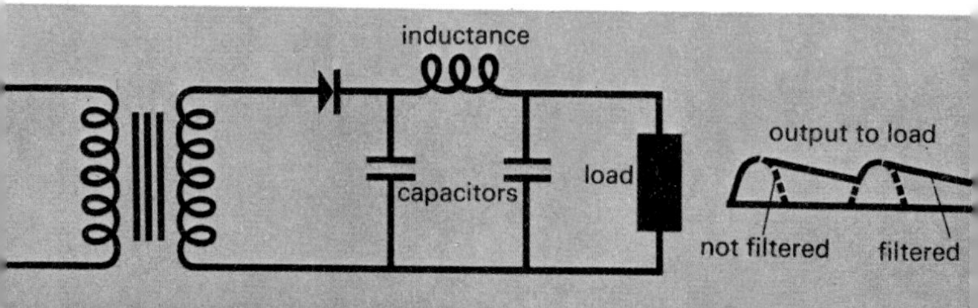

when it is not and so current is available continuously.

All the rectifiers described so far are free running, that is, they conduct when the supply voltage on the anode is positive with respect to the cathode. If they can be set so that they only conduct when initiated from a separate source, they are known as *controlled rectifiers*. This is very useful, for example for speed control of motors, since the amount of d.c. power can be controlled from zero to full output.

This control is applied mainly to mercury arc rectifiers, although *thyristors*, which are double junction solid state units (silicon controlled rectifiers), are now available and are coming into major use. In the former, the inter-electrode gap is increased until the arc will not form without the help of an auxiliary trigger electrode, the firing time of which can be set to any point on the half cycle.

Controlled rectifiers can be used to convert d.c. to a.c., for running portable power tools from a battery where no mains supply is available. They are then known as *inverters* and, by feeding pulses of d.c. at a given repetition rate (controlled from a separate source) into an inductance and capacitance circuit, a sine wave can be produced.

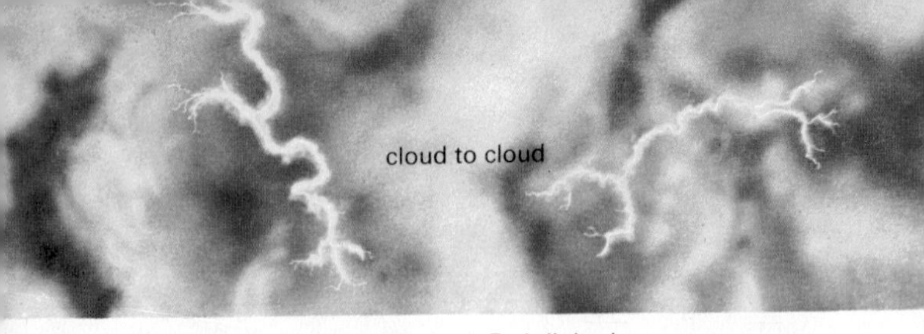

(*Above*) Sheet lightning. (*Below*) Fork lightning

Atmospheric electricity

Thunderstorms provide the most spectacular displays of natural electricity. Over the years scientists had speculated on the cause, but it was Benjamin Franklin (page 8) who first showed that it was a form of static electricity.

A lightning discharge occurs when enough electric charges of opposite polarity have accumulated and separated in the clouds, causing a high potential difference between the cloud and earth or between different parts of the cloud. When the electric field becomes high enough, electrons are detached from the molecules of air, which then suddenly 'breaks down' (loses its insulating properties) and charge flows through the lightning stroke to equalize (or discharge) the two regions of opposite polarity. The thunder is the noise produced by this sudden release of energy. Thus the time difference between the flash and the bang allows the distance from the observer to the stroke to be estimated. The speed of sound in air is about 0·2 miles per second.

A cloud to earth discharge appears as a jagged line and is called fork lightning, whereas a cloud to cloud discharge, although still fork lightning, is usually seen as a glow in the cloud, due to diffusion effects in the cloud. This is referred to as sheet lightning.

Measurements taken from aircraft have shown that the charge distribution in clouds is substantially as shown. How this occurs is not fully understood as yet, but one of the more widely accepted explanations is as follows. The charges are caused by spray electrification, assisted by the strong upward air currents at the head of the storm cloud. Moisture is carried up to the cooler regions, where it condenses into drops which, when they attain sufficient size so that gravity overcomes the upward air currents, fall. As they fall they break up into smaller drops, releasing negative charge, and so becoming positively charged. These small, now positively charged, drops are carried up by the air currents leaving a net negative charge in the lower part of the cloud and a positive one at the top. Some of the positively charged drops fall as rain. Thus the rain at the front of the cloud is positive and that at the rear is negative.

The potential differences so built up are of the order of 100 million to 1000 million volts, and currents of up to 500,000 amperes have been recorded in lightning strokes. Thus the energy dissipated in one lightning stroke is greater than the output of a large modern power station.

Typical charge distribution in a thunderstorm

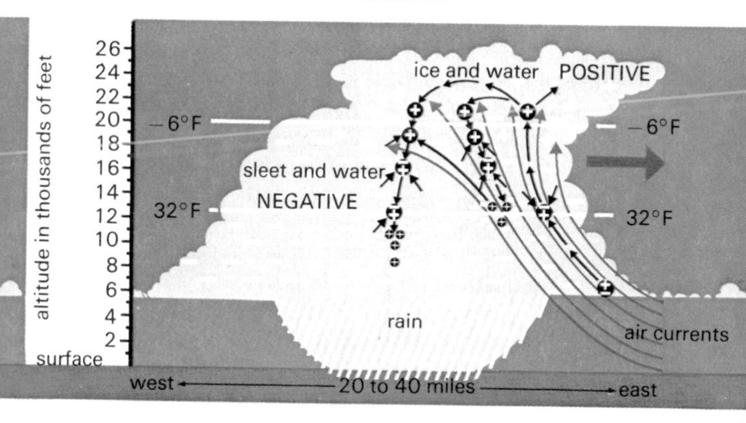

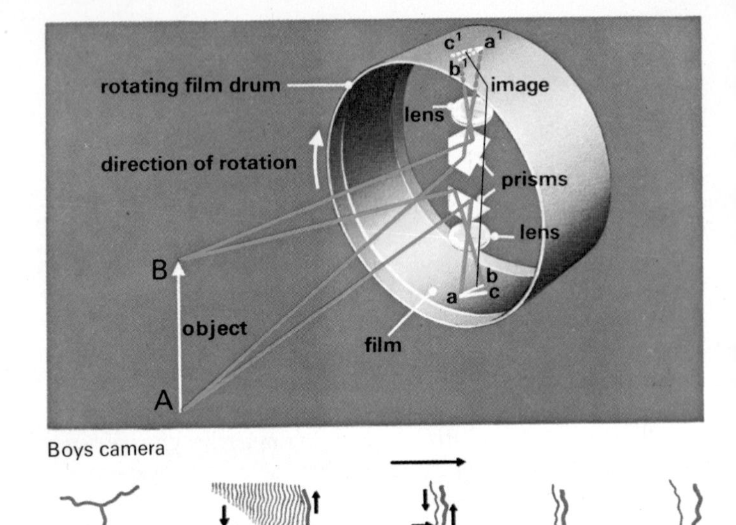

Boys camera

Cloud-earth lightning discharge with normal camera (*first left*) and Boys camera

The nature of the lightning stroke has been fairly well documented over the years from both field observations and laboratory investigations. One of the most useful investigating tools has been the *Boys* camera, developed in 1900 by C. V. Boys. With the drum stationary, the light rays from an object moving over the path from A to B will pass through the lower prism and trace path a-b on the film, and also path a′-b′ via the upper prism. If the drum is revolving (usually about 6500 rpm) the paths will be distorted, being from a to c and a′ to c′. A comparison of the two images will reveal the amount of distortion, which is a measure of the velocity of the moving object as the drum rotational speed is known.

Each lightning stroke to ground actually consists of a series of separate discharges occurring so quickly that to the eye, or normal camera, only one is evident. Each discharge has a faintly luminous *leader* from cloud to ground (not visible to the eye) followed by the intensely luminous return main stroke from ground to cloud moving at a much higher velocity, back along the initial path. The leader velocity is 4×10^7 cm/sec and the main stroke velocity is 4×10^9 cm/sec (speed of light is 30×10^9 cm/sec). The leader to the first discharge has different characteristics to those of the following discharges. It consists of a series of streamers moving downwards, over the same path, in a step by step manner. Hence it is called a *stepped* leader. Each step is about 50 metres in length, and the time between them is about 100 microseconds. Subsequent steps extend the path in slightly different directions, since they seek the nearest highly charged area to discharge. Thus when the return stroke follows the initial path of the stepped leader, the characteristic zig-zag pattern results. Subsequent leaders are of a dart-like character and follow the path of the initial stroke, this being almost a conductor, since many free electrons are left following the passage of the initial main stroke. These subsequent strokes are caused by different areas of the cloud discharging sequentially.

Another instrument used for studying lightning strokes was the klydonograph. This was especially useful before the actual shape of the electric fields in a discharge could be looked at by using an oscilloscope. The klydonograph consisted of an electrode mounted on a photographic plate insulated from the ground. This electrode was attached to the place, usually a transmission line, that was liable to be struck by lightning. Each polarity produced a different pattern, called a Lichtenberg figure, and the magnitude of the stroke could be estimated from the area of the patterns.

(*Left*) Klydonograph. (*Right*) Lichtenberg figures

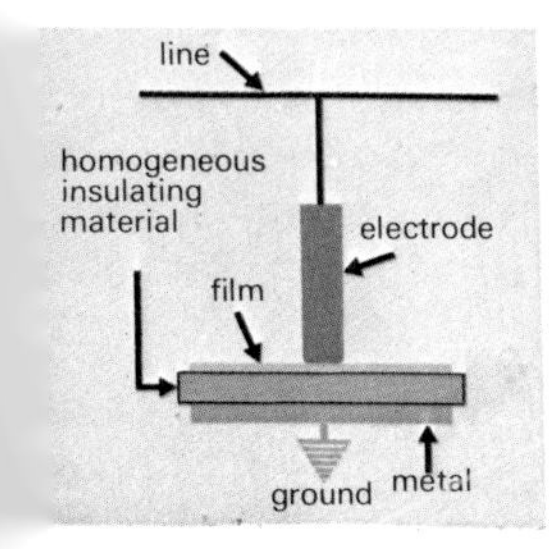

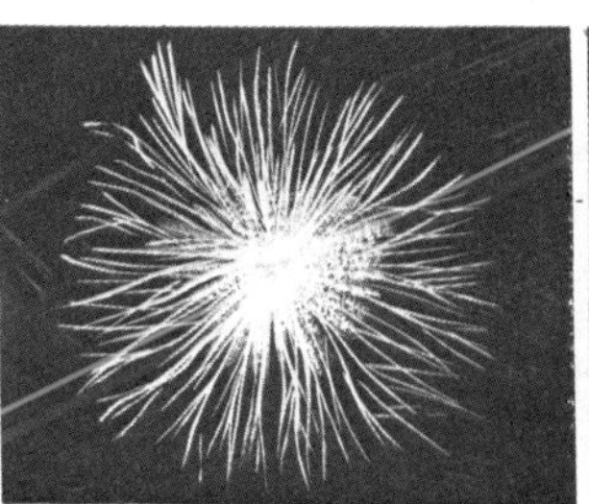
POSITIVE

NEGATIVE

Aurora borealis

St Elmo's fire

Apart from causing forest fires, one of the most devastating effects of lightning is its disruption of the electricity supply system. Transmission lines are prone to lightning strikes, since they are in effect sharp points on flat ground and, as such, set up points of high stress concentration. When lightning strikes a transmission line, tower, or the ground nearby, large voltages appear across the insulators, overstressing them and causing a flash-over. This takes the line out of service, probably damaging the insulators permanently in the process. This subject will be discussed further in a later section.

Buildings are often struck by lightning, causing fires

and often much damage. Protection against this is obtained from lightning rods, which are conductors placed vertically or horizontally on roofs of buildings and which are solidly connected to earth. When these conductors are struck the energy passes harmlessly to earth without causing damage to the structure.

One of the more pleasant manifestations of atmospheric electricity is the aurora borealis or 'northern lights'. This phenomenon, which is visible in the extreme northern and southern latitudes, is a spectacular and very beautiful display of moving colours. The glow spreads across the sky at a height of between 50 and 250 miles. At this height the air pressure is so low that charged particles that have been emitted by the sun, cause it to ionize, or glow, in a similar manner to a glow discharge lamp (page 123). These charged particles are deflected by the Earth's magnetic field and become concentrated at the magnetic poles, and hence the phenomenon is generally visible only in extreme northern and southern areas of the world.

St Elmo's fire is another form of visible atmospheric electricity. It occurs mainly on the masts of ships and mainly in pre-storm conditions in the tropics. The charged atmosphere near the storm causes small electric discharges, called *corona*, to occur on sharp points. This phenomenon can be stationary or can move around. In the past sailors were often very superstitious and this phenomenon was interpreted as a visitation from heaven and hence named after the patron saint of sailors, St Elmo.

Flash-over caused by lightning

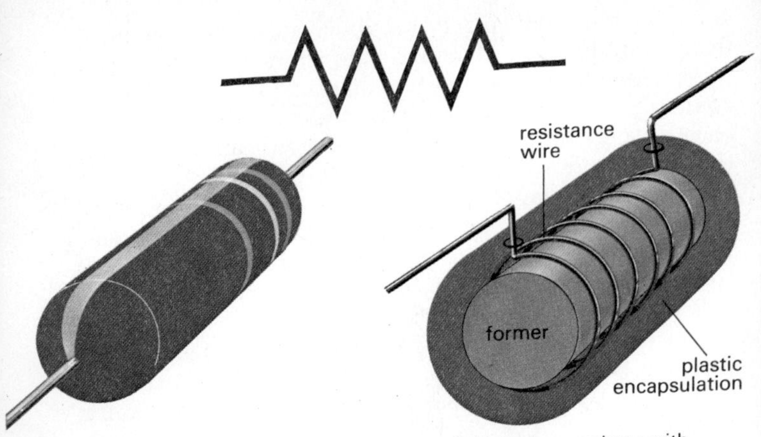

(*Top*) Resistor circuit symbol. (*Bottom left*) Carbon resistor with colour code for identification. (*Bottom right*) Wire-wound resistor

ELECTRIC CIRCUITS

An electric circuit is any path, or combination of paths through which a current flows. When current is flowing, physical characteristics of the circuit tend to oppose the flow, and this is known as the *impedance* of the circuit. Impedance is composed of three elements, the resistance, the inductance and the capacitance of the circuit.

Resistance

Resistors are elements that offer physical resistance to the current flow and are the only ones that use power. All materials, including conductors, have some resistance. Usually it is deliberately added to a circuit to achieve certain conditions (see the companion volume, *Electronics*).

For small currents the solid carbon resistors are the most suitable. Wire-wound resistors are bulkier but can withstand high currents. As the resistance of a wire is proportional to its length and inversely proportional to its cross-section, a large range of resistances can be constructed from wires of varying diameter and length. Carbon resistors are suitable for a.c. and d.c. use, but wire-wound must be used with care on a.c. due to the *inductance* of the wire coil.

Inductance

When a.c. flows through a coil of wire, a changing magnetic field is set up which links the turns of the coil and induces a counter current which opposes (due to Lenz's law) the initial current. This counter current is smaller than the initial current, since the magnetic coupling is poor, and a resultant current flows. This magnetic resistance to current flow is *inductive reactance* (X_L) and the property of the coil causing it is the *self inductance* (L). This effect depends on a changing magnetic field, and therefore on the frequency (f) of the current change. The inductance depends only on the dimensions of the coil. Thus $X_L = 2\pi fL$. Inductance has no effect on d.c., since the magnetic field is unchanging.

In circuits, inductance is usually a parasitic effect due to the use of coils and wire-wound resistors etc., but in some circumstances it is required (page 97). Inductances (or *chokes*) are just coils of wire, the value being increased by the use of an iron core which concentrates the magnetic linkage, with consequent higher reactance.

The unit of inductance is the *henry*, and that of inductive reactance is the *ohm* as it is a resistance to current flow. One ohm of inductive reactance passes a current of 1 ampere when energized with 1 volt.

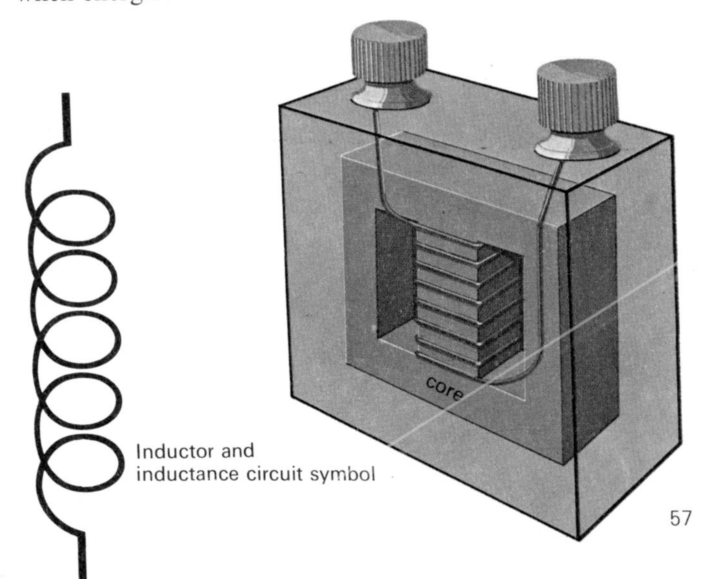

Inductor and inductance circuit symbol

Hydraulic analogy of a capacitor

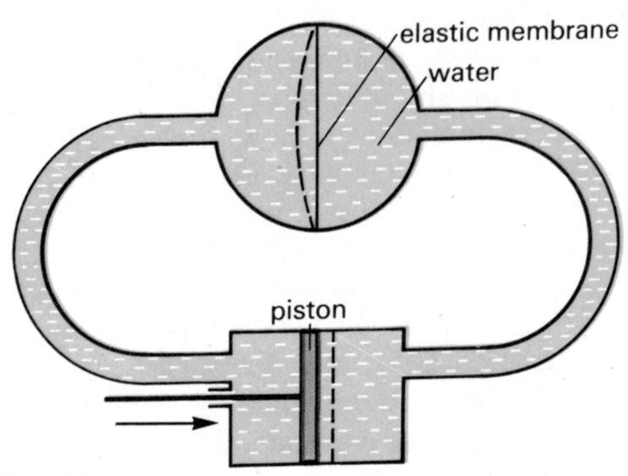

Capacitance

Under certain conditions electricity can be stored (page 8) in a circuit and the devices which do this are *capacitors*. These consist of flat metal plates separated by insulation but having no direct conducting path through the assembly. Modern capacitors are long sheets rolled like a swiss roll. Storage of electricity (or electric charge) can be likened to filling a container with a gas: the higher the pressure the more gas it will hold and so the higher the voltage the more charge can be stored. The charge (q) is thus proportional to the applied voltage (V) and the capacitance (C), which is the ability of the capacitor to store the charge, that is $q = CV$. Now charge flow is an electric current, and therefore V/q is an impedance called *capacitative reactance* (X_c). Again, with a.c. the voltage is constantly changing and so the reactance depends also on the frequency and $X_c = \frac{1}{2\pi fC}$.

The current in a capacitor is found to lead the voltage by 90°, because when fully charged (at maximum voltage) no more charge can flow (thus current is zero) and when discharged (zero voltage) the maximum charge (and thus the current) can flow and the current flows before the voltage.

For d.c., capacitors act as an open circuit since, once charged, no more charge (thus current) can flow. With a.c. the capacitor is being alternately charged and discharged, and thus a current is always flowing, even though there is no physical connection through the unit. The electricity does not flow through the insulator, but electrons are transferred from one plate to the other around the external circuit, by the electrostatic field set up between the plates. In a hydraulic analogy, the circular chamber represents the capacitor, the piston the applied voltage and the water the current. Shifting the piston to the right (d.c. case) causes the water in the right-hand side to flow, stretching the elastic membrane and thus causing the water in the left-hand side to move until equilibrium is reached and motion ceases. If the piston is continually moved back and forth (a.c. case), there will be an alternating flow of water. The flow acts as if it were not bounded by the elastic membrane, which is what happens to the current in the a.c. case.

The value of the capacitance depends only on the physical dimensions of the unit, and on the charge storing property of the insulation, that is, its *permittivity*. This latter is the ability to concentrate the electrostatic field between the plates, air, for example, having a value of unity, while porcelain and glass have one of about six. Thus the higher the value the larger the capacitance. Capacitance is measured in *farads* (F), which are large units, the most common values being in the microfarad range. Ohms are the units of capacitive reactance and again, 1 ohm of capacitative reactance passes a current of 1 ampere when energized with 1 volt.

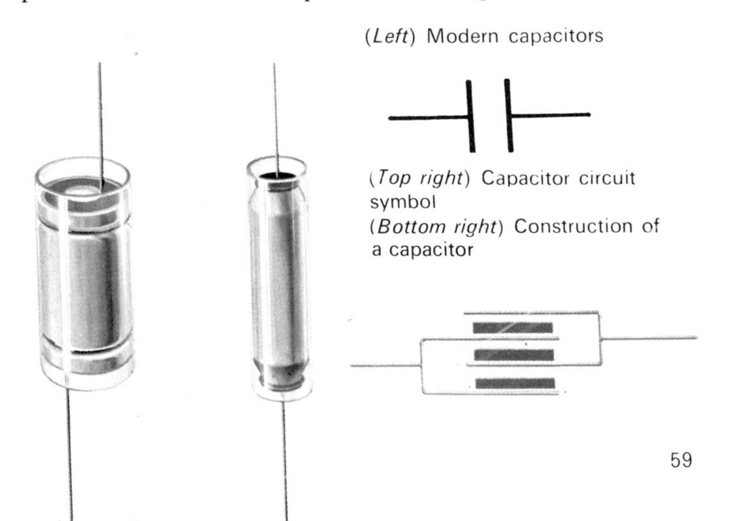

(*Left*) Modern capacitors

(*Top right*) Capacitor circuit symbol

(*Bottom right*) Construction of a capacitor

Phase relationships of a.c. circuits

Resistance

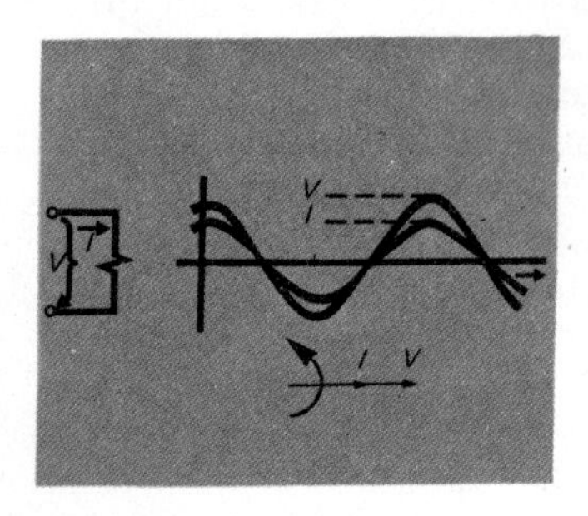

Inductance

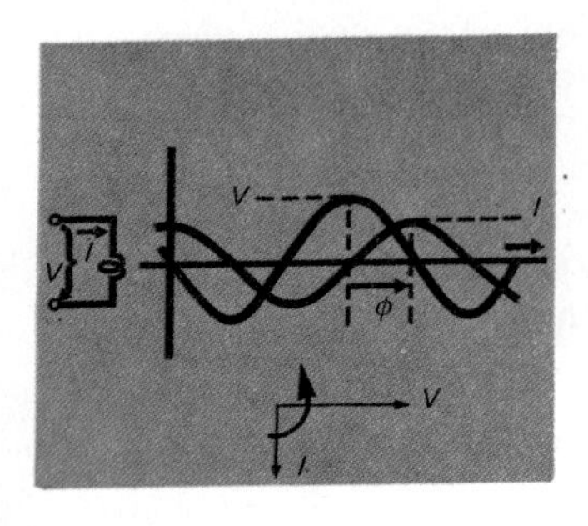

Capacitance

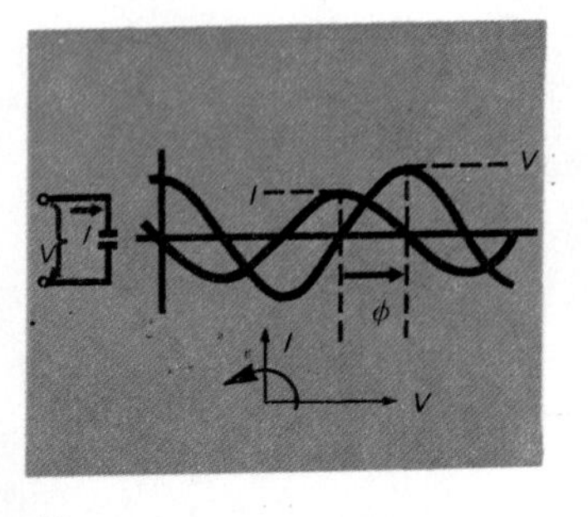

The phase relationship of voltages and currents must be accounted for when their values are calculated for a.c. circuits. Ohm's law must be generalized to include the *impedance* (Z) in place of the resistance, that is, $I = \frac{V}{Z}$.

For d.c. and purely resistive a.c. circuits the impedance is the resistance. For combinations of resistance, inductance and capacitance, the impedance is a combination of the resistance and the reactance (inductive reactance being 90° behind the resistance and capacitative reactance 90° ahead), and the current is not displaced a full 90° but an amount related to the value of resistance included in the circuit. Impedance is found by adding the resistance and reactance in a way that accounts for the 90° displacement between them. This is *vector* addition and is based on the fact that the square of the hypotenuse of a right angle triangle is equal to the sum of the squares of the other two sides. Thus $Z^2 = R^2 + X^2$ and $Z = \sqrt{(R^2 + X^2)}$ and the angle of the impedance is related to the value of $\frac{X}{R}$, this being

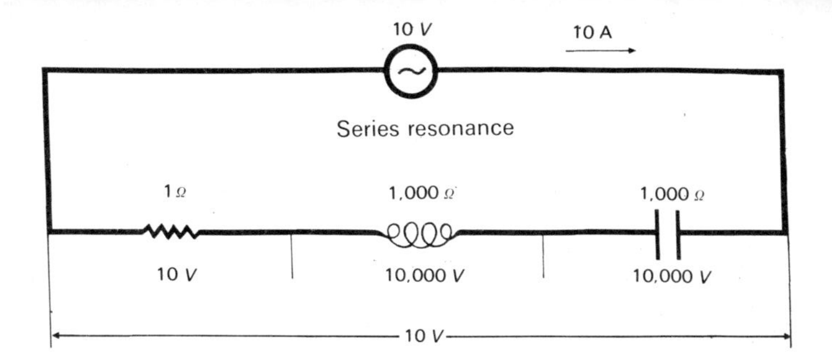

the angle that the current is displaced from the voltage. The ratio $\frac{X}{R}$ is the tangent of the angle and values of the angle can be found from mathematics handbooks which list the tangents for all angles. This angle must always be specified to define the exact value of an impedance. If incorporated in the same circuit, X_L and X_c tend to cancel each other and as by definition X_c is negative, $Z = \sqrt{(R^2 + (X_L - X_c)^2)}$.

Resonance

This interesting phenomenon occurs when the inductive and capacitative reactance are equal (which is only possible at one frequency). The supply only 'sees' the resistance in the circuit and large currents can flow if the elements are connected in series and very high voltages then appear across each reactance. For example, a 1Ω resistor in series with 1000Ω of inductive reactance and 1000Ω of capacitative reactance will draw 10 amperes from a 10 volt source, and a voltage of 10,000 volts will appear across each reactive element. However, as these are in opposition, one being 90° ahead of the current, the other 90° behind, the voltage across both elements together is zero. This circuit is used to generate high voltages for television sets, where little power is required.

If the reactances are in parallel, they still cancel each other and at resonance draw no current from the supply, although large currents circulate around the $L-C$ loop, each supplying the other with reactive power on alternate portions of the cycle. This is used to overcome inductance and capacitance in power transmission systems (page 97).

bushing
tank
core
windings

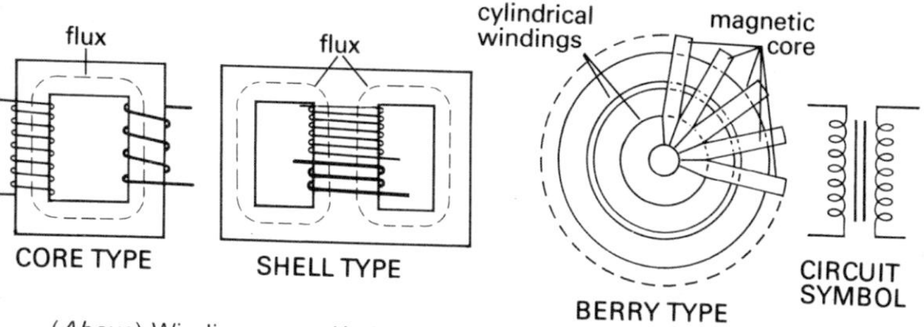

(*Above*) Winding types. (*Left*) Power transformer, 400 kV, 1,000 MVA. (*Left inset*) Shell type transformer. (*Right inset*) Radio transformer

Transformers

The transformer is the basis of all power transmission. It is basically two inductances in close proximity. A current in one coil sets up a magnetic field which induces a current in the second coil. As current can only be induced by a changing magnetic field, transformers only work with a.c.

The magnetic field strength depends on the number of turns on the primary, or input, coil (n_p) and the current flowing (I_p), expressed in ampere-turns. As the ampere-turns in the core will be constant for both coils, the output current (I_s) can be varied by changing the number of turns on the output, or secondary coil (n_s). Thus $I_p n_p = I_s n_s$. The energy taken out of the secondary cannot exceed the energy fed into the primary coil. Thus $W_p = W_s$. Using Ohm's law we find $V_s = (n_s/n_p)V_p$, which shows how a transformer can raise or lower voltage, the amount depending only on the ratio of the turns in the primary and secondary coils.

In practice the output voltage and current is slightly lower than the turns ratio predicts, due to heating losses in the conductors and *eddy* current losses. The latter are due to currents, induced in the iron core, which generate heat but do not contribute to the output.

Power transformers have a laminated iron core, consisting of several thin cores in parallel and insulated electrically to reduce the area available for eddy currents to circulate without reducing the magnetic effectiveness. The shape of the core varies and upon it are wound two, suitably insulated coils. This assembly is placed in an oil-filled tank, the oil acting as insulating and cooling medium, with suitable terminations for connection into the system.

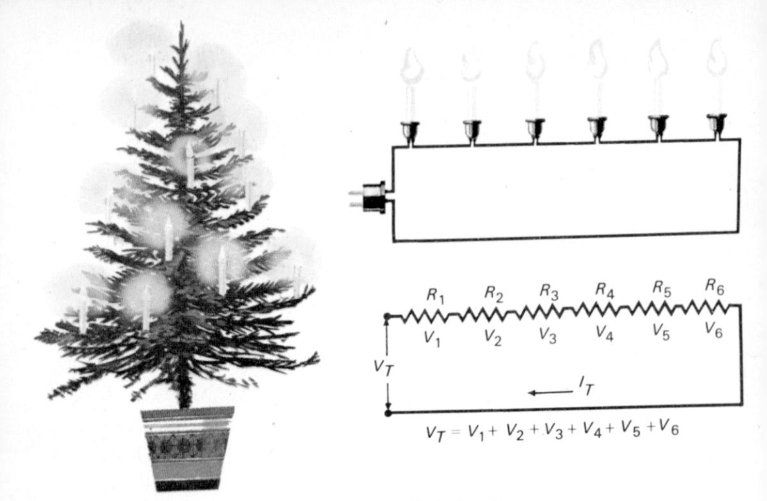

(*Above*) Series circuit. (*Below*) Parallel circuit

Circuit elements (resistors, inductors and capacitors) can be connected in two basic ways, either in series or in parallel. In a series circuit all the components are connected end to end and the same current flows through each element, the voltage dividing itself between the elements in proportion to their impedance. Parallel circuits have each end of the components connected to a common point, the voltage across each element thus being constant and the total current being the sum of the currents in each element.

When circuits have their elements connected in series or parallel, or a combination of both, it is useful to be able to determine a total impedance (Z_T) for the circuit so that the total current drawn from the source may be calculated. For series circuits, the total impedance is the sum of the individual impedance, that is, $Z_T = Z_1 + Z_2 + Z_3$ etc. Thus resistances and inductances are totalled, $R_T = R_1 + R_2 + R_3$

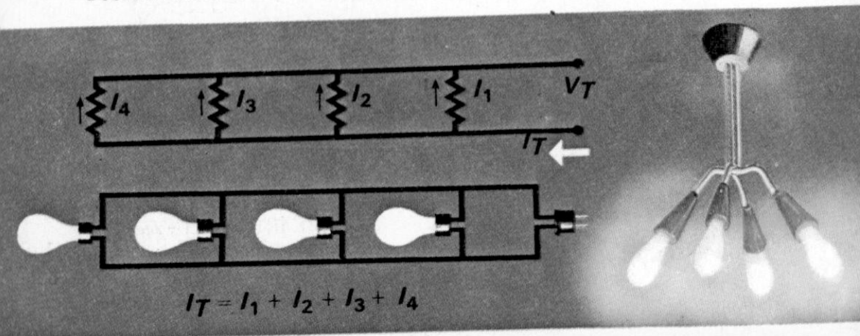

and $L_T = L_1 + L_2 + L_3$ (using vector addition as required). As capacitative reactance is the reciprocal of capacitance, $\frac{1}{C_T} = \frac{1}{C_1} + \frac{1}{C_2} + \frac{1}{C_3}$, so that capacitors in series have a total value lower than the individual values. In a parallel circuit the total current is the sum of the currents in each element, $I_T = I_1 + I_2 + I_3 = \frac{V_T}{Z_T} = \frac{V_T}{Z_1} + \frac{V_T}{Z_2} + \frac{V_T}{Z_3}$. Thus $\frac{1}{Z_T} = \frac{1}{Z_1} + \frac{1}{Z_2} + \frac{1}{Z_3}$. Resistance and inductances thus have a total impedance smaller than any individual value, but for capacitors the total value is the sum of individual ones.

By using these fundamental relationships, almost any complex circuit can be reduced to an *equivalent* or total impedance. A simple example, using only resistors is illustrated. Firstly, the 2 and the 4 ohm resistors are in series so that the total is 6 ohms, thus leaving a parallel circuit of a 6 and a 3 ohm resistor (a). This can be reduced to an equivalent resistance of 2 ohms (b) thus the total resistance seen by the 6 volt battery is 2 ohms (c) and it supplies 3 amperes to the circuit.

Parallel circuits are the most common in general use, since most devices run off a common voltage source (the mains) each device drawing the amount of current it requires. Series circuits find specialized applications in electronics, but a common use is in Christmas tree lights where ten, cheap, 24 volt bulbs are connected in series to work off the 240 volt mains. However, if one bulb fails, all go out and no indication of which one is faulty is evident, whereas in a parallel circuit only the faulty light fails.

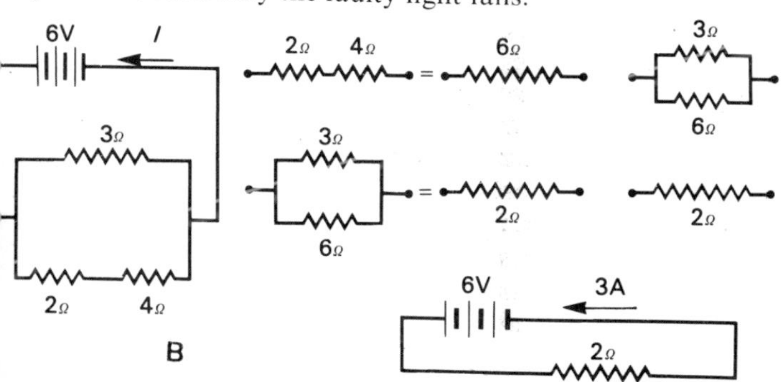

Reduction of complex circuits

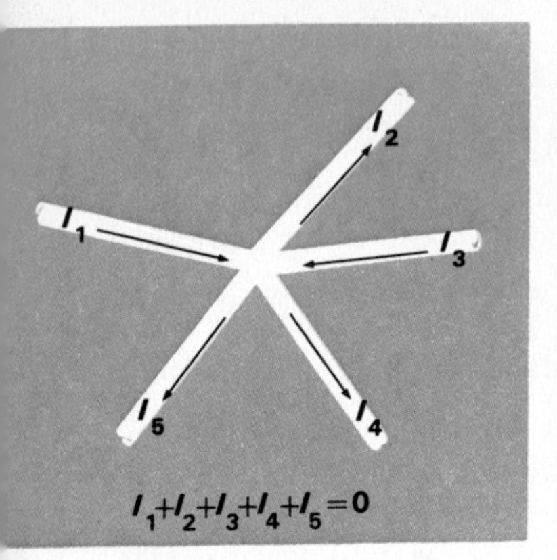

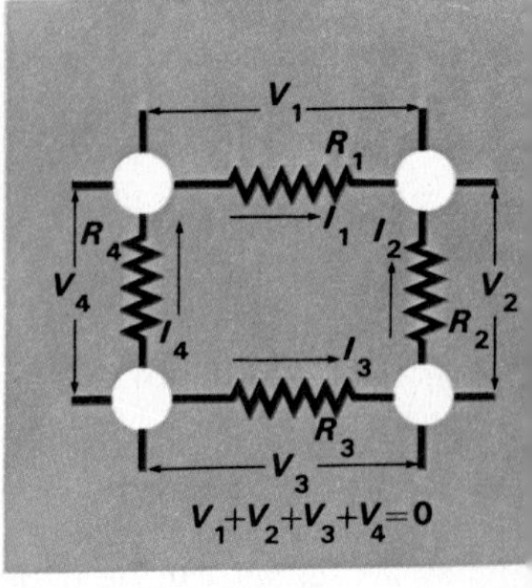

(*Left*) Kirchoff's current law. (*Right*) Kirchoff's voltage law

Very complex circuits are tedious, and in some cases too complicated, to solve by straight resolution into a single impedance, using the series and parallel circuit reduction rules. It is also usually necessary to know the current flowing through a component somewhere in the middle of the circuit. A number of circuit theorems have been devised of which Kirchoff's laws are the most versatile. The laws are simple: (1) the sum of the currents at any junction of conductors is zero and (2) the sum of the voltage drops around a closed loop is zero. Voltage drops are the voltages across components due to the currents flowing through them (for example, one ampere through ten ohms give a voltage drop of ten volts) or the magnitude of any voltage sources in the closed loop. These laws are obvious after a little thought and form the backbone of circuit analysis. To demonstrate these laws the circuit on the previous page will be solved using each in turn.

Voltage law

Two loops are present and so two currents, I_1 and I_2, which cause the voltage drops, are assumed to be flowing. Anticlockwise rotation is assumed to be positive, and if the current value turns out to be negative the wrong direction of flow was initially assumed. In loop 1, $I_1R_1 - I_2R_2 - V = 0$, (I_1 and I_2 flow in opposite directions in R_1, and therefore the voltage drop is the difference of the drops caused by each current), and in loop 2, $I_2(R_1+R_2+R_3)-I_1R_1=0$. Substituting the values of resistance and voltage, $3I_1-3I_2=6$ and $9I_2-3I_1=0$. From the latter, $I_2=\frac{1}{3}I_1$ and when this is substituted in the former, $3I_1-3(\frac{1}{3}I_1)=6$ and $I_1=3\text{A}$. This solution of the equations is a simple application of the technique known as *simultaneous equations*. Neither equation can be solved on its own as both I_1 and I_2 are unknown but together the answer can be obtained. There must be as many equations as there are unknown currents and this could be a large number in complex networks and so computers are often used to solve this type of problem.

Current law

There are two junctions, A and B, but in this particular case the currents are the same at both and so only A need be considered. The law gives the equation $I_1+I_2-I_3=0$ (the sign indicates assumed direction of current flow, negative being into the junction, and if the final answer has a negative sign it just means that the wrong direction of flow was assumed initially). By inspection $I_1=\frac{V}{R_1}=\frac{6}{3}=2\text{A}$; $I_2=\frac{V}{R_2+R_3}=\frac{6}{2+4}=1\text{A}$. Thus $I_3=I_1+I_2=2+1=3\text{A}$. This

Circuit solution using voltage law (*left*) and current law (*right*)

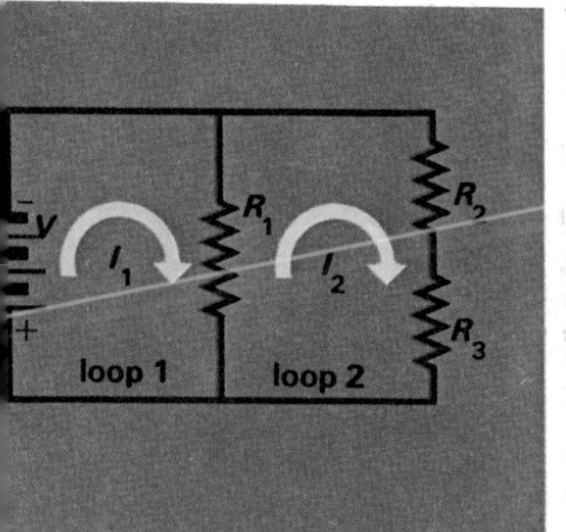

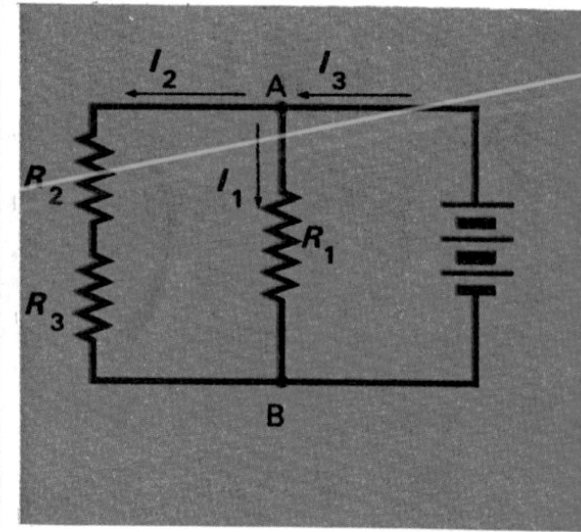

is a very simple case and normally the currents could not be determined so easily, and so the method is not shown to its best advantage. In a large network many junctions would occur, each one giving an equation, and the simultaneous equation technique would be used, or again a computer if it were very large.

Power in an a.c. circuit is still the voltage multiplied by the current, but instantaneous values must be used and the total power determined by averaging them over a cycle. In a purely resistive circuit, the current and voltage are in phase and the power is always positive (although the direction of current flow has reversed, it still dissipates heat), and is the product of the *effective* voltage and current, $W = VI$.

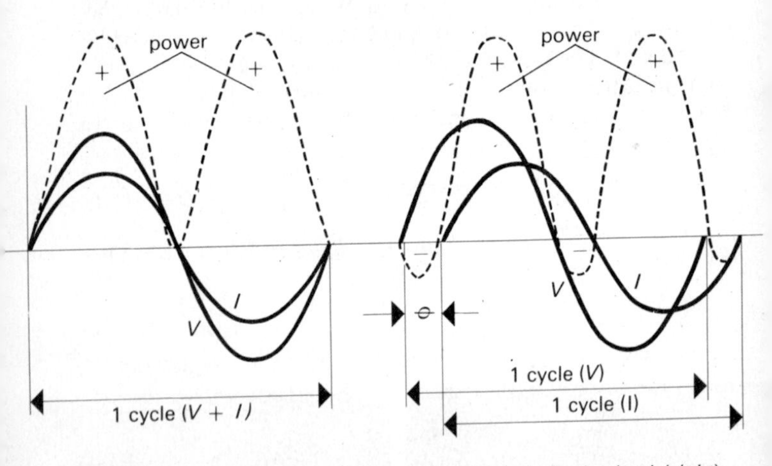

Power in a resistive load (*left*) and a resistive-inductive load (*right*)

For purely inductive circuits, the power is a sine wave at twice the original frequency, having equal positive and negative components flowing in opposite directions and the total power is zero! A capacitative circuit behaves similarly with the curve being displaced one quarter of a voltage cycle. Thus reactances absorb and store power (an inductance in its magnetic field, a capacitance in its electrostatic field) for part of the cycle and then return it to the source during the remainder of the cycle. Pure reactances therefore do not cause power dissipation in a.c. circuits. This power exchanged between the reactances and the source is called *reactive power*.

In practice resistance is present in all circuits (all wires have some resistance), and resistive-inductive circuits, where the current lags the voltage by less than 90°, are most common. Here the output power is more positive than negative over a cycle and the average or *active* power is then $W = VI \cos \theta$. The term $\cos \theta$ is the cosine of the angle between the voltage and current, which is known as the *power factor* (PF) of the circuit. It is always less than unity and is the ratio between the *average* and the *apparent* power. The latter is the product of the effective values of voltage and current and is important since, although reactances dissipate no power, they pass current and the source must supply this current for part of the cycle. Most a.c. equipment is usually rated by its apparent power, called volt-amperes (VA), and a power factor instead of the real power in watts; the exception being pure resistance loads, for example, heaters and irons.

As an example, a 240 V, 2·4 kVA, universal motor (page 111), with a PF of 0·75 would draw 10 amperes from the mains and have an output of 1·8 kW. On d.c. the same motor would take 10 amperes but have an output of 2·4 kW. An electric fire, say 1 kW, would draw $\simeq$4 amperes on a.c. or d.c.

In a.c. circuits a wattmeter indicates the average power (page 74) and so by measuring the voltage and current as well, the power factor of a circuit and thus the impedance and its angle can be determined, that is, $\cos \theta = \dfrac{W}{V \times I}$

As most industrial uses involve inductive-resistive loads the power factor can be low. Supply authorities do not like this since they supply the apparent power but are only paid for the average power and so they set minimum values for the power factor. Low values are improved by adding capacitance in parallel to cancel the inductance and 'correct' the power factor. The source supplies active power, the reactances supplying each other with reactive power.

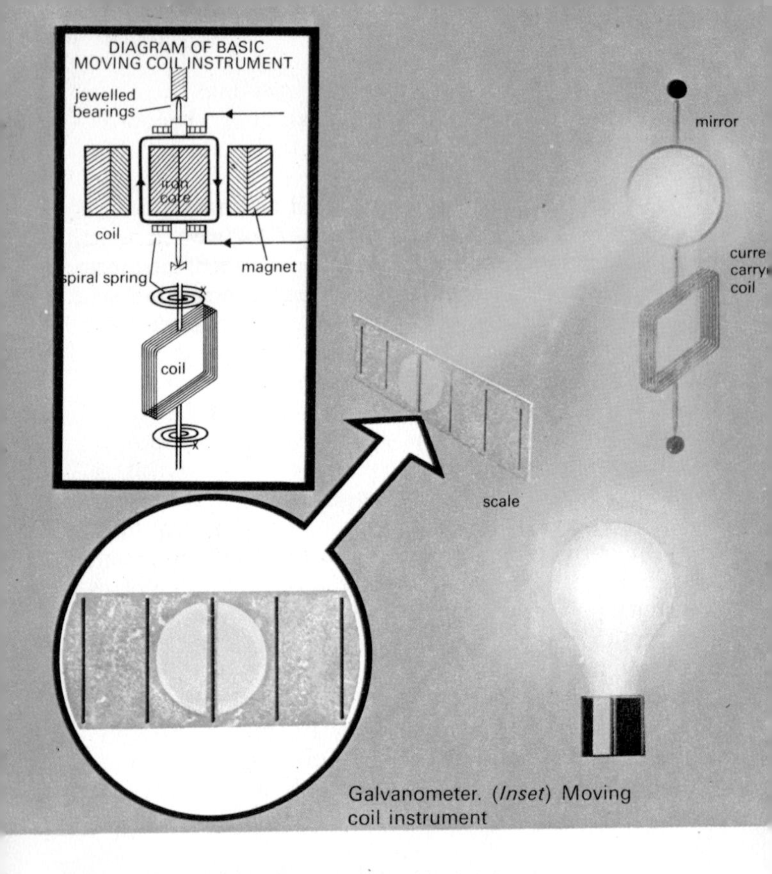

Galvanometer. (*Inset*) Moving coil instrument

ELECTRICAL MEASUREMENTS

Electrical measurement is vital to electrical engineering, since we must know what is happening in an electrical system. As there is no obvious sign of the presence of electricity, use must be made of associated phenomena, the heat generated or the magnetic fields set up. This latter phenomenon formed the basis of the earliest method and most modern instruments are still dependent on it.

Current measuring instruments

The *galvanometer* is the basic instrument for measuring very small direct currents (less than $10\mu A$). It consists of a small coil, which carries the current to be measured, suspended so that it can turn on its vertical axis inside a fixed magnet. The magnetic field set up in the coil causes it to rotate until it is in equilibrium with the restraining force, due to the twisting of the suspension wires. A mirror is mounted on the coil and a light beam is reflected on to a scale to indicate the amount of deflection.

Moving coil instruments are based on the galvanometer but are much more robust and are used to measure higher currents (up to 5A). They are still limited to d.c. use, since they require a steady magnetic field in the coil to give a deflection.

The coil is mounted in jewelled bearings and small spiral springs tend to restore it to its zero position and also to act as current leads. A pointer is fixed to the coil and the deflection is read off a scale. Operationally, the coil current sets up a torque which is resisted by the spiral springs and the coil takes up a position where the driving force is equal to the restoring force. The driving force is directly proportional to the applied current and thus the deflection is linear and indicates the average current flowing.

For direct measurement of a.c. a different type of instrument must be used. This *moving iron* type has a fixed coil and a pivoted vane of soft iron (to which the pointer is attached), which is attracted by the magnetic field set up by the current in the coil. The operation of this instrument depends on the energy stored in the magnetic field and the deflection is therefore proportional to the energy involved and therefore to the power and thus to the square of the current. This type therefore has a *square law* scale and thus a direct reading of the effective value of the current is obtained.

Moving iron instruments are cheap and rugged, but not as accurate as moving coil types, and so they are used as portable instruments for field work or where great accuracy is not needed. They also read d.c. directly but to less accuracy. Larger currents (up to 300A) can be measured directly as the current does not have to flow through the spiral springs.

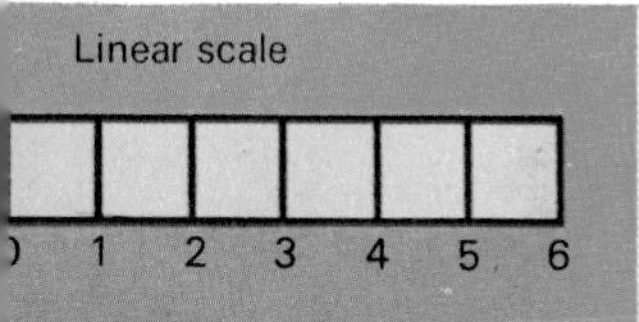

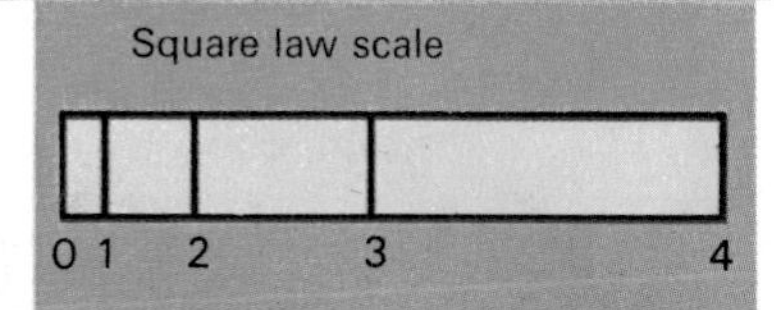

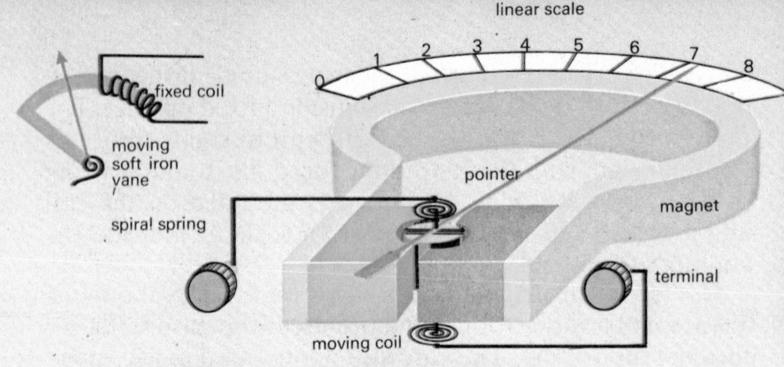

(*Left*) Moving iron ammeter. (*Right*) Moving coil ammeter

By using rectifiers, the moving coil instrument can measure a.c., and the mounting of small solid state rectifiers directly into it has made it today's standard a.c. measuring instrument. However, it only indicates the average current and the indicated deflection must be scaled by a constant factor in order to read the effective value. The instrument can still be used for d.c., but the scaling factor must be ignored.

To extend the range of these instruments so that higher currents can be measured, low resistances called *shunts* are connected across the coil. The resistance is chosen so that only a small proportion of the current flows through the instrument, the rest bypassing it through the shunt. Since the shunt and the instrument resistance are known, the factor by which the instrument reading must be multiplied to obtain the total current flow can be calculated.

Shunts can be used to extend the measuring range of moving iron instruments but current transformers (page 79) are usually used, since for a.c. the inductance of the shunt must be allowed for and non-inductive shunts are difficult to design for operation with large currents.

Rectifier instruments

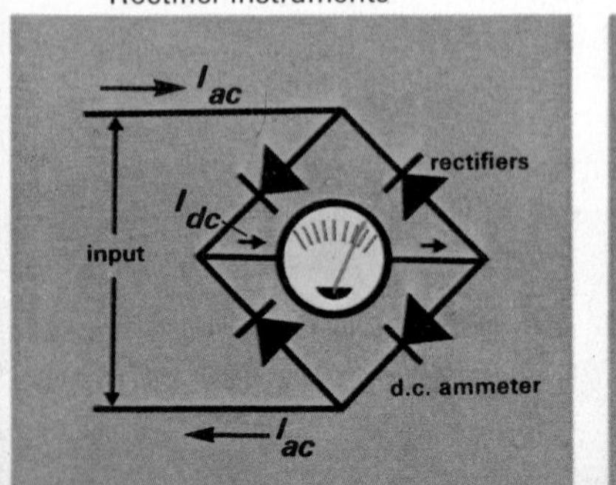

Electrothermic instruments

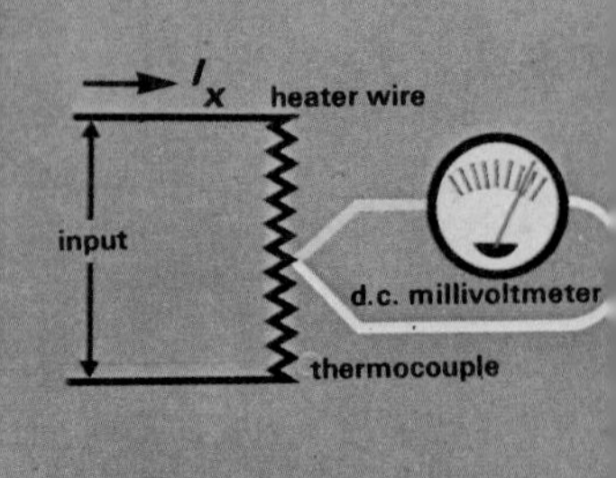

Electrothermic instruments do not depend on the magnetic effect of the electric current but on the heating effect. The main uses are for measuring very high frequency a.c. (up to 1 MHz), where moving coil and iron instruments are inaccurate due to eddy current losses, and as *transfer* instruments for comparing a.c. and d.c. instruments. This latter use is possible because, by definition, the average value of the d.c. and the effective value of a.c. generate the same amount of heat.

These instruments comprise a heater wire with a thermocouple at its centre. The thermocouple output (d.c.) can be accurately measured. For highly accurate transfer measurements the ambient temperature must be known.

Voltage measuring instruments

All the above instruments are fundamentally current measuring devices. By inserting a high resistance of known value in series with them, voltages can be measured by applying Ohm's law. The scales of the instrument are directly calibrated in voltage.

Multi-range instruments

For many uses it is convenient to be able to measure a wide range of voltages and currents without having to use many instruments. By mounting a rectified moving coil instrument in a case and including suitable shunts and series resistances, which can be selected by switches, a *multimeter*, capable of measuring from 0·1 milliamperes to 10 amperes and from 2·5 to 1000 volts a.c. and d.c., can be produced. The a.c. scaling factor can be allowed for in a.c. resistance and shunt values and thus one scale can be used for both a.c. and d.c.

Use of shunt

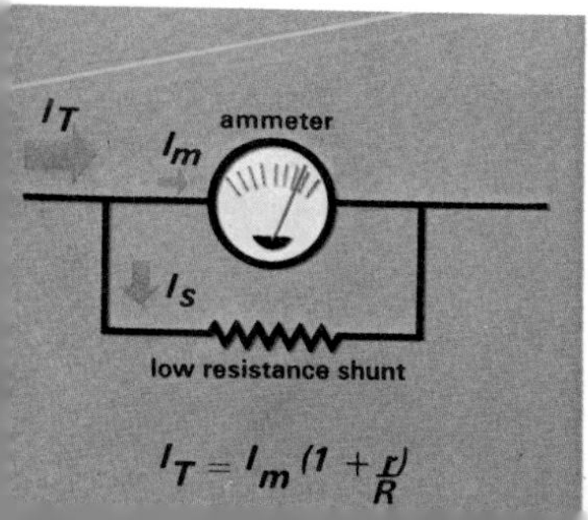

I_T ammeter I_m I_s low resistance shunt

$$I_T = I_m\left(1 + \frac{r}{R}\right)$$

Voltmeter construction

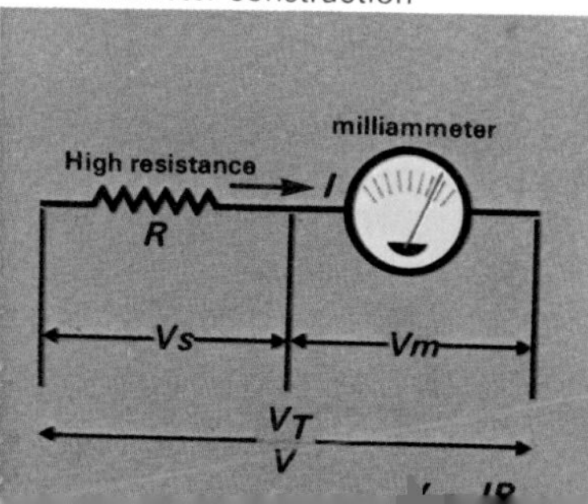

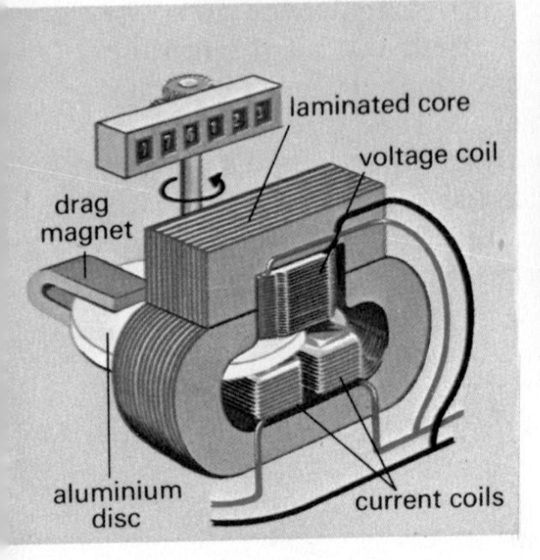

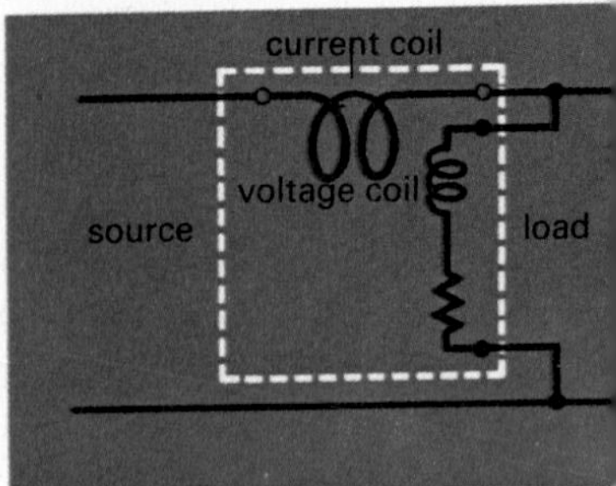

(*Left*) Internal view of kilowatt hour meter. (*Right*) Wattmeter circuit diagram.

Power measuring instruments

Wattmeters measure the power being used in a circuit at any given moment. This equals the product of the applied voltage and the current flowing (page 28). For d.c. an ammeter and a voltmeter measure these values but for a.c. this is not generally possible, due to the phase difference between the current and the voltage which must be accounted for (page 69) as a purely resistive circuit, which has no phase difference, is unusual in practice.

The wattmeter consists of two coils, the current coil, having few turns and being fixed, and the voltage coil, having many turns and being pivoted with a pointer attached. This is like a moving coil instrument (page 70) with the magnet replaced by the fixed coil. The resulting torque, which deflects the pointer, is the product of the currents in the two windings. As these carry the instantaneous voltage and current in their proper time relation, the phase difference is automatically accounted for.

To get an accurate reading the voltage coil must be connected on the load side of the wattmeter, otherwise the voltage drop due to the current coil will be registered as part of the power being consumed. The instrument can be used with voltage and current transformers to extend its range, the indicated reading being multiplied by the transformer ratios.

To measure the total power used in a given period a watt-hour meter must be used. In these induction meters an aluminium disc rotates, due to motor action set up by the phase difference between the magnetic fields of the current and voltage coils. If the voltage and current are in phase, the current and thus the magnetic field of the voltage coil is 90° out of phase, because the coil is virtually a pure inductance (page 60). So the disc is subjected to a moving magnetic field (page 108). For a pure inductive load, the current is already 90° out of phase and as the voltage coil current is again shifted by this amount, the magnetic field becomes in phase and no rotation takes place. The speed of rotation is proportional to the phase angle between voltage and current and to their magnitudes and thus to the average power. Eddy currents induced by motion of the disc past the permanent magnet ensure that the disc stops immediately the power is switched off. A train of gears connect the disc axle to dials which indicate the number of revolutions, and, as the number of revolutions per kilowatt hour is known, the total energy consumed is registered. By suitable gearing the dials are made to read directly in kilowatt hours.

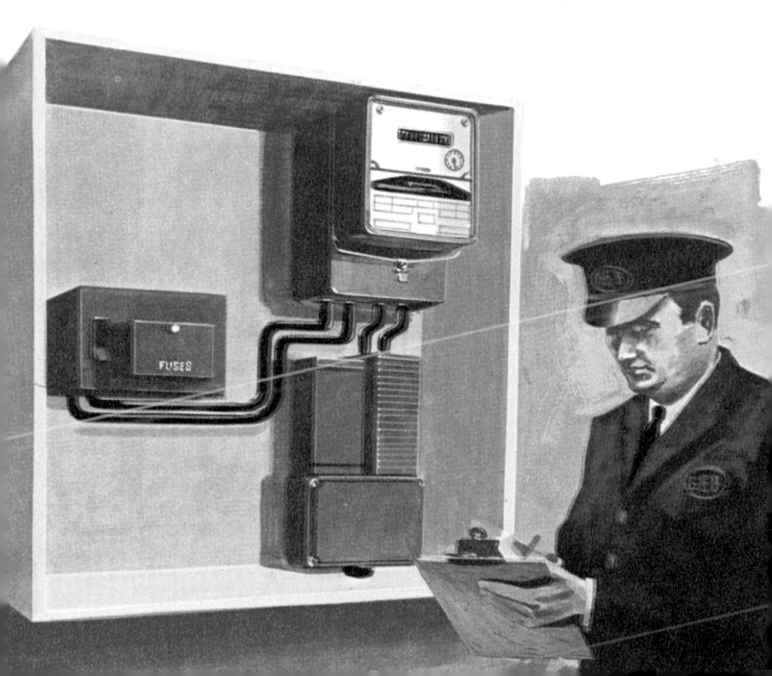

Bridge circuits

Special *bridge circuits* have been developed for accurately determining the values of resistors, capacitors and inductors. The value of the unknown component is found in terms of an accurately known standard unit. In the basic bridge circuit Z_x is the unknown, Z_s is the standard and Z_1 and Z_2 are variable impedances of known value. Each impedance is known as an *arm* of the bridge, and the latter two are balancing arms. The method of measurement is known as a *null* method because the arms Z_1 and Z_2 are varied so that the voltages at points B and D are identical, the galvanometer (null detector) thus indicating a lack of current in the *bridge*. Under this balance condition $i_{AD}Z_x = i_{AB}Z_s$ and $i_{CD}Z_1 = i_{CB}Z_2$ and thus $Z_x = \frac{Z_s Z_1}{Z_2}$

Three important types of bridge are described below.

Wheatstone bridge

This circuit is used, with a d.c. supply, for determining resistance values. R_s is a known fixed resistor and R_1 and R_2 are variable over a wide range. At balance $R_x = R_s \frac{R_1}{R_2}$.

Schering bridge

This circuit is used for measuring the capacitance and resistance of a capacitor. In practice all capacitors have resistance although it is usually very high so that they are effectively open circuits to d.c. Capacitor resistance is of particular importance in insulation testing, where it is expressed as the dielectric loss angle (DLA). The lower this value the better the quality of the insulation. The circuit consists of an a.c. source (of frequency f), a capacitance standard, a variable capacitor (decade type) and the usual resistors in the balancing arms. At balance $C_x = C_s \frac{R_1}{R_2}$, DLA $= 2\pi f C_2 R_2$.

Owen bridge

Inductance (L) is measured in terms of a capacitance standard using this circuit. This avoids the use of variable inductors which are difficult to construct accurately. An a.c. source is used and the balance arms are C_4 and R_4. Being composed of wire coils, inductors have resistance which is measured at the same time. At balance $L = C_1 R_2 R_4$, $R = \frac{C_1}{C_4} R_2$.

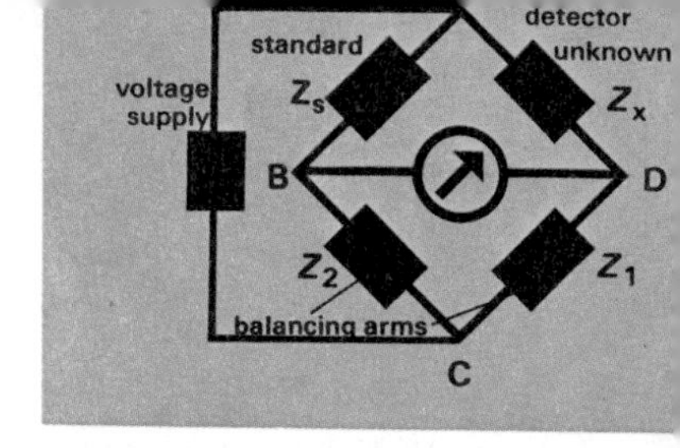

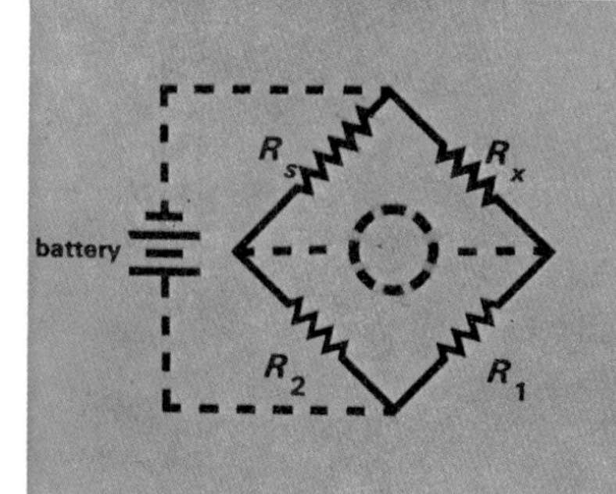

(*From top to bottom*) General bridge circuit, Wheatstone bridge, Schering bridge, Owen bridge

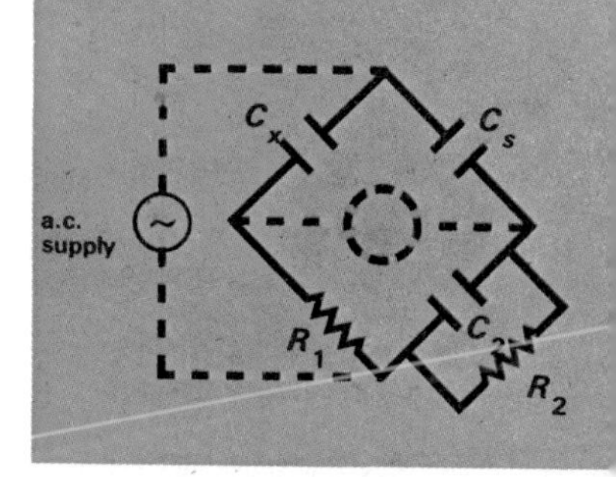

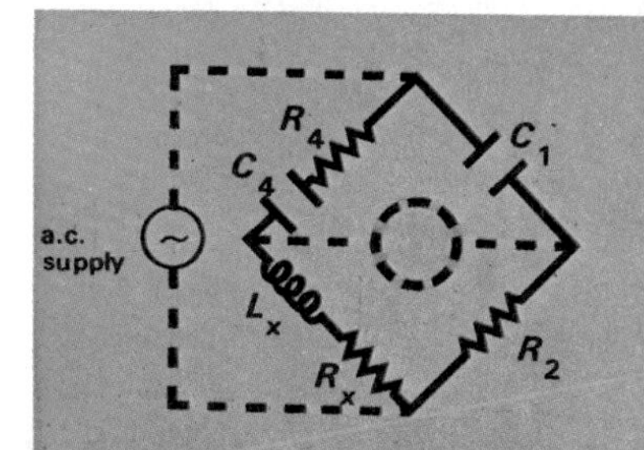

Miscellaneous measuring techniques

Insulation limitations and safety requirements prohibit the use of instruments for directly measuring voltages greater than about 2500V. Above this voltage, dividers are used to tap off a small portion of the voltage which can then be read on a normal, suitably scaled instrument. These devices consist of a high (Z_h) and low impedance (Z_l) in series and a voltmeter, connected as shown. To measure 10kV an impedance of 20kΩ and one of 200Ω would be used with a voltmeter capable of reading $\frac{10{,}000 \times 200}{20{,}000 + 200} = 100\text{V}$, but with the scale marked from 0 to 10kV instead of 0 to 100V. Capacitors and resistors are the usual impedances, the latter for d.c. and low alternating voltages and the former for high a.c. voltages, as it is easier to make high voltage capacitors than resistors.

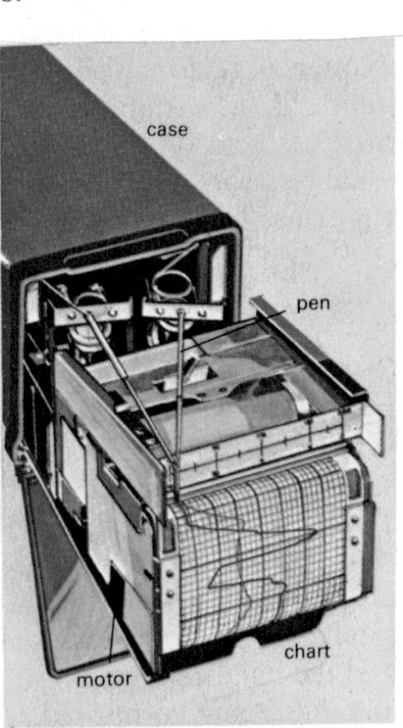

(*Above left*) Voltage divider. (*Above*) Chart recorder. (*Left*) Voltage divider circuit. (*Far right*) Voltage measuring transformer. (*Right*) Current measuring transformer

Transformers are used to measure high alternating currents and voltages, since their ratios can be accurately determined. A voltage transformer has many turns on the primary (n_p) and few on the secondary (n_s) while a current transformer has a single turn primary, usually the cable passing through the coil centre, and a large number on the secondary.

Thus $V_m = \frac{n_s}{n_p} V_a$ and $I_m = \frac{n_p}{n_s} I_a$.

It is sometimes necessary to know how a measurement varies with time and to avoid having to continually watch a meter, recording instruments are used. In the simplest version a chart is driven by a motor at a fixed speed and the pen is attached to the pointer of a moving coil type of instrument movement, and thus a continuous record results. This type of instrument takes a few seconds to respond to any change and is thus used for slowly changing measurements.

For fast changes, reflecting instruments (page 70) (with ultra-violet light sources) are used in conjunction with a sensitized chart which develops when exposed to daylight.

Modern measuring techniques make use of sophisticated electronic instruments which are more sensitive and accurate than the normal types. The cathode ray oscilloscope and the valve voltmeter are two that come to mind and these, and many others are described in a companion volume to this book, *Electronics*.

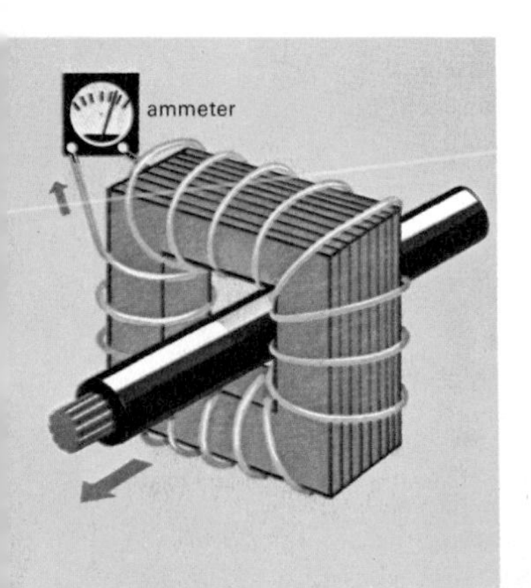

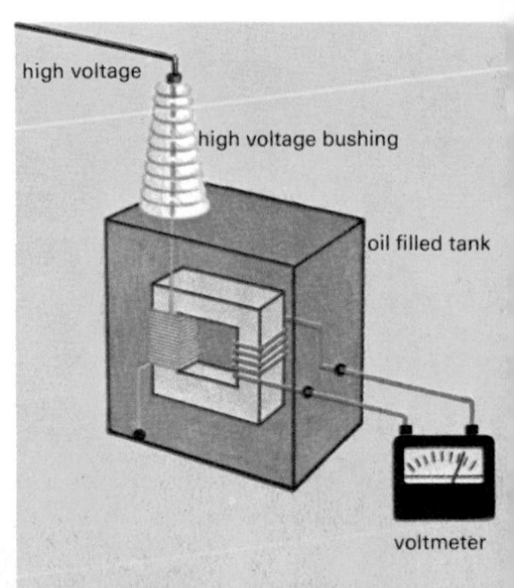

dam
generator
penstock
turbine
water flow

MODERN POWER GENERATION

Hydroelectric power stations

This is one of the cheapest ways to generate electricity, the kinetic and potential energy of running water being harnessed to drive the generators. However, it is limited to mountainous regions where fast rivers flow. A fairly large difference in height is required to provide enough force in the flowing water to turn the water-wheels (turbines) to which the generators are connected.

Early installations had a vertical water-wheel turning the machinery. Modern plants use vertical generators mounted on horizontal water-wheels. These generators, of the synchronous salient pole type, are of very large diameter (30 to 40 feet), and have a large number of poles (up to 60) depending on the speed of rotation (usually from 100 to 600 rpm).

The required head of water is obtained by constructing a dam across the river, usually at its steepest point to make maximum use of height. A reservoir forms behind the dam, so that a constant head of water is obtained. This stored water is used in times of drought to power the turbines.

The water from the reservoir is carried through underground ducts (penstocks) and then directed against the turbines. The generators are usually mounted in the *powerhouse*, situated below the dam.

Electricity is generated at voltages within the range 6·6 to 15kV. The *exciter* set on top of the main generator provides the low voltage, high current d.c. required to set up the magnetic field in the main generator. Transformers, situated just outside the power house raise the voltage to the long distance transmission levels of 275 and 400kV.

These plants are cheap to run, with only a small staff needed. However, the initial capital required is enormous, since dam construction is a major civil engineering feat. As suitable sites for hydroelectric stations are seldom near built-up areas, the cost of transmission lines must be added. However, these stations have a long life, of up to eighty years, compared to twenty for a thermal station.

Hydroelectric power station and cross-section of dam and powerhouse

Thermal power station flow diagram

Thermal power stations

Thermal power stations, where the heat energy of a fuel is converted into electrical energy, provide the majority of electricity in use today. Their main advantage is that they can be sited where needed thus saving on the cost of transmission lines. However the high cost of fuel (coal, oil or gas) is a disadvantage.

In operation fuel is burnt to heat the water in the boiler, which then produces steam at very high pressure (up to 5000 lb/square inch) and temperature (1100°F). This steam is directed on to the blades of a high speed turbine, which is connected to the generator. The steam is then condensed and the water returned to the boiler, thus forming a closed cycle.

This condensation requires either large amounts of external cooling water (15 million gallons per hour for a 300MW station) or large cooling towers. Thus sites must be chosen near a river or where the unsightly cooling towers will not detract from the surroundings.

The generators are of two pole construction rotating at 3000 rpm (steam turbines are more economical to run at high speeds). Again the voltage output varies between 6·6 and 33kV, the larger the machine rating the higher the voltage used. The exciter set is again mounted on the shaft. At present 500MW sets, 23,000A at 22kV, are the largest in service but 1000MW sets are being designed.

Large amounts of heat are produced, due to the high currents generated, and so cooling of the generators is important. Fans mounted at each end of the rotor circulate air or hydrogen gas. Hydrogen is a better cooling medium than air and also has only one fourteenth its density. Thus less power is required to circulate it. Recently, liquid cooling using oil or water has been adopted. Liquids are far superior to gases for cooling, water being 50 times better than air.

Thermal stations are not particularly efficient, the overall figure being about 38%, that is, only about 38% of the thermal energy put into the plant from the fuel is available as electric energy. The main limitation is the boiler design, the generators being up to 90% efficient.

Experiments are now in progress to run these stations by computer and thus reduce the large staff requirement.

Nuclear power stations

Nuclear power stations are thermal stations where the heat is obtained from a nuclear reaction instead of burning fuel.

Radioactive elements, such as uranium and plutonium, have unstable nuclei which emit neutrons that strike other molecules, splitting them up, thus releasing more neutrons and producing heat. The process continues and an uncontrolled chain reaction results. Moderator materials, carbon or heavy water, absorb neutrons without their nucleus splitting. So by mixing the two types of material the chain reaction is controlled at a constant level of heat output.

The heart of the station is the nuclear reactor, which consists of a central container holding the 'fuel' (radioactive material). Rods of moderator material can be inserted or withdrawn from the fuel element, withdrawal initiating the reaction, the rate of which is governed by the amount of withdrawal. The heat generated is carried by a coolant (gaseous carbon dioxide, water or liquid sodium) from the core to the heat exchanger, where the steam for the turbine is produced. The *Boiling Water Reactor* (BWR) has no heat exchanger, the water being evaporated into steam inside the reactor core. The core is shielded to prevent harmful radiation from escaping.

Stringent safety precautions are observed and all operations are *fail safe*, that is if anything goes wrong the complete process shuts down. A special set of control rods, called *scram rods*, are poised outside the core ready to enter at high speed. They are gravity operated, being held by an electromagnetic clutch, and can shut down a plant within a

Nuclear power station flow diagram

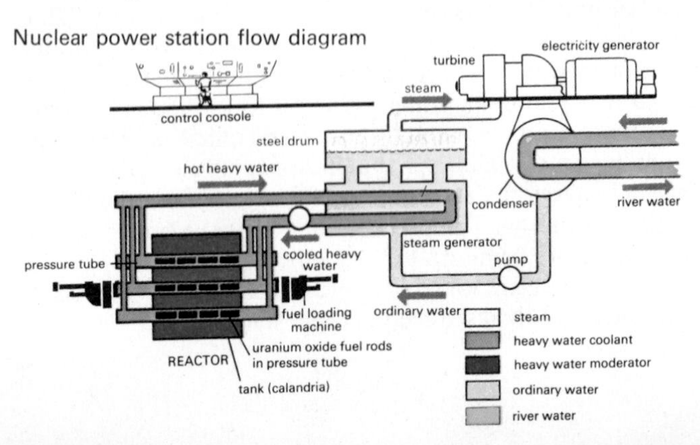

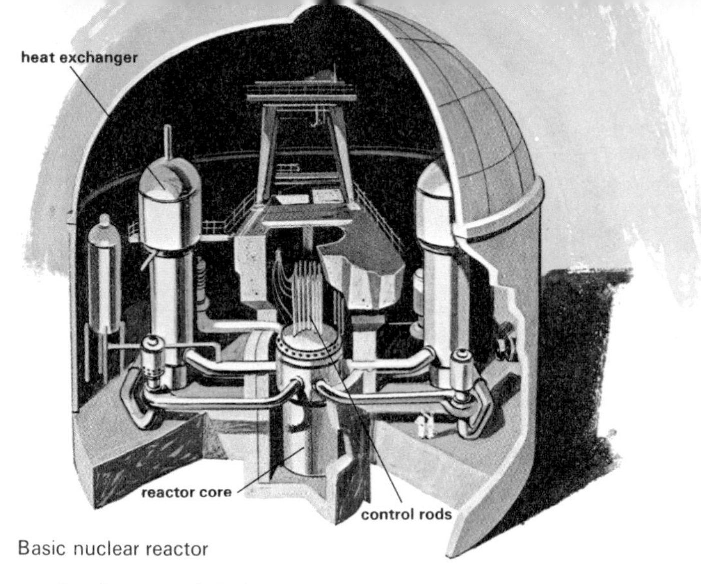

Basic nuclear reactor

tenth of a second. With increasing confidence nuclear plants are now being built near populated centres.

In Britain there are nine nuclear power stations of the *Magnox* gas cooled design with three of the *Advanced Gas Cooled Reactor* (AGCR) design due for completion in 1972. A number of small UKAEA (United Kingdom Atomic Energy Authority) experimental stations are also in operation. The new stations will be of 1200MW size and 41% efficiency rather than the 500MW size and 28% efficiency of present ones. They will also be cheaper than present ones, whose generating cost is higher than that of conventional stations although no fuel need be supplied after the initial filling. The *breeder* reactors produce more fuel than they consume and this can fuel other reactors.

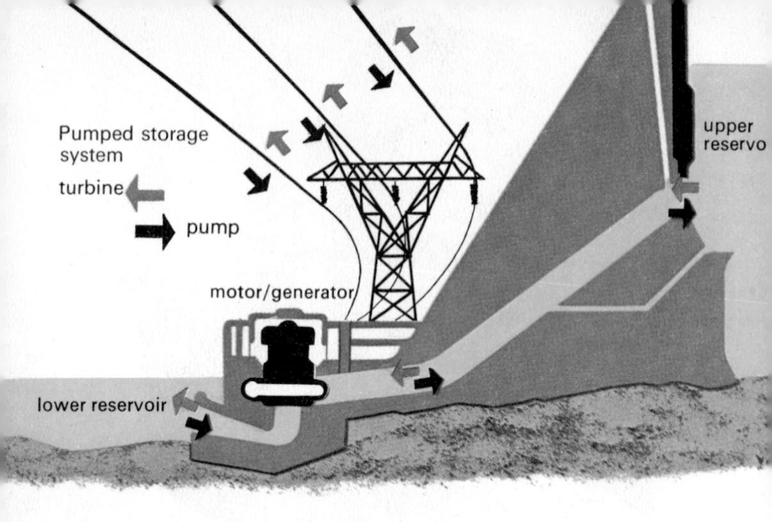

Miscellaneous methods

Variations on the hydroelectric generating system are tidal and pumped storage systems. Large amounts of water are stored in reservoirs during slack periods, for example, at night, and then used to generate electricity when needed. In the pumped storage scheme the turbines pump water into the upper reservoir in slack periods. Tidal systems depend on the ebb and flow of the tide to fill the reservoir.

High speed generators, driven by gas turbines (basically aircraft engines) supply limited amounts of power in emergencies or peak periods. Electricity can be generated within minutes, whereas a conventional station can take up to twenty-four hours to become operational.

When certain materials, mainly selenium and silicon, are exposed to light, electrons are liberated and a current flows in an external circuit. Power levels are low and sunlight is the main source of light. This *photoelectric* effect forms the basis of *solar cells*, which supply power in remote areas and satellites and charge batteries.

Similar to the thermocouple principle (page 36) some semiconductors, such as germanium and silicon, liberate electrons when heated and thus generate electricity. This *thermoelectric* effect is increased by connecting the elements thermally in parallel but electrically in series, like a thermopile. These generators have no moving parts and are extremely robust. A typical use is the powering of a radio (valve type) from the waste heat of an oil lamp. Small radioactive heat sources have been used and also heat from the Sun.

A *fuel cell*, such as a *Bacon* cell, obtains electricity, by chemical action, direct from fuels without using a turbine/generator. Its action is similar to a battery, except that the electrodes (compressed nickel powder) are not consumed and it will function indefinitely as long as fuel is provided. Hydrogen molecules give up two electrons each at one electrode (which becomes the cathode) and these can flow through the load. The hydrogen ion then passes through the electrolyte (normal hydrogen or electrons are unable to) and when it reaches the other electrode it passes through, gaining two electrons and leaving a positive charge on the electrode which thus becomes the anode. The hydrogen is removed by combining with the oxygen to form water, which is pumped away. Alcohol can be used as a fuel but hydrogen is best. Fuel cells are the most efficient method (up to 80%) yet devised of producing electricity and large powers can be produced. However, the high operating temperature, about 240°C, makes cooling necessary. These units are still in the developmental stage but have been used successfully in spacecraft (page 133).

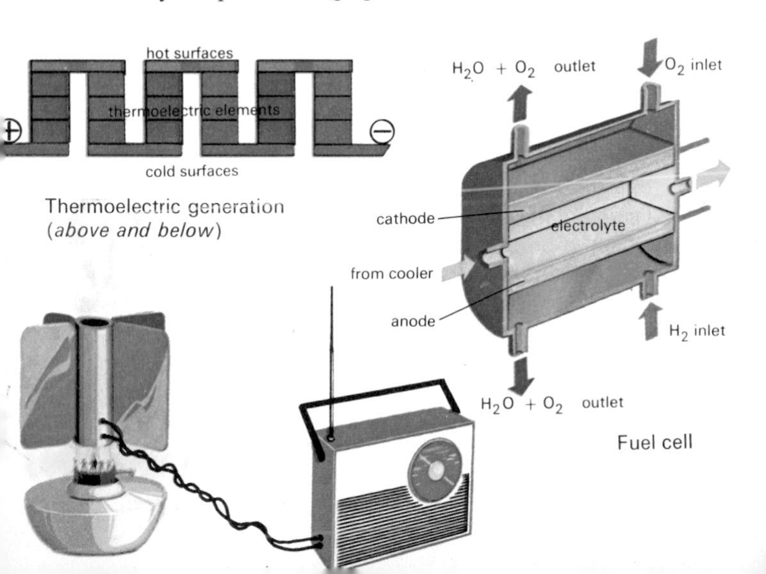

Thermoelectric generation (*above and below*)

Fuel cell

ELECTRIC POWER TRANSMISSION

One of the great advantages of electricity is the ease with which it can be transferred from one place to another. This transfer of electricity is known as *electric power transmission* and the interconnection of wires and apparatus that carries it out is known as a *power system*.

Power systems have three principle components; the generating stations (for producing electricity), the transmission system (for transmitting large quantities of electricity to the area where it is needed) and the distribution system (for distributing the electricity at lower voltages to individual consumers).

Alternating current is used in virtually all modern power transmission systems because of its ease of transformation from one voltage to another, and the simpler construction of generators and motors using it.

Efficiency is the paramount factor in power system design and operation. The system must transmit power to where it is needed without losing it along the way. As the amount of power being transmitted increases (modern systems are now transmitting up to 2250MW, enough to work over 2 million electric fires) the transmission voltages must increase in order to keep the efficiency high. Power loss is dependent on the I^2R loss of the line. Thus the higher the voltage the lower the current ($P = VI$) and the lower the losses. The highest voltage in the British Isles is 400kV but some overseas countries have 735kV systems. High voltages bring additional problems in insulation requirements and corona loss. Corona is the blue glow surrounding the conductors of high voltage lines at night. It is due to the high electric field causing the air to light up in the same manner as a neon sign. This requires power and will thus reduce the amount of power available at the end of the line.

Transmission lines operate at voltages of 132kV and above, while distribution systems operate at lower voltages (66, 33 and 11kV in the United Kingdom). Consumers are normally supplied at 240V single phase or 415V three phase. Except for North America and parts of Japan where 60Hz is used, the standard frequency of a.c. systems is 50Hz.

1A52

Three phase systems

Power transmission today is invariably a three phase system. All transmission towers carry multiples of three conductors, double circuit lines of six conductors being the most common. Three phase operation is preferred as it makes more efficient use of the generators, has a steady rather than pulsating power flow (thus motors run more smoothly) and is more economical in conductor utilization.

What is a three phase system? In an idealized generator armature (see below) a coil (shown in red) of three turns has its active sides in opposite slots. The coil could be extended on either side but this would be inefficient. More efficient generation is obtained by fitting another coil (yellow) in the slots at an angle to the original one and finally filling the rest of the slots with a third coil (blue). Thus, there are three separate windings or *phases* acting as three separate generators in a single machine. It will be seen that each phase rises to a peak in turn and that there is always a voltage difference between each phase. (Because the phases are at an angle of 120° to each other the voltage between each phase is 1·73 times the voltage across each phase.)

These three phases can be connected in various ways. They may function as three independent supplies to loads in various places using six wires, or, if the loads are in the same area, a common return wire, which carries the sum of the three currents, can be used and only four wires are required. What size return wire is needed? Three times the size would be the initial thought, but with equal (or balanced loads the currents will be balanced and the sum of these is zero! Thus the fourth wire is unnecessary.

Three phase generator and voltage waveforms

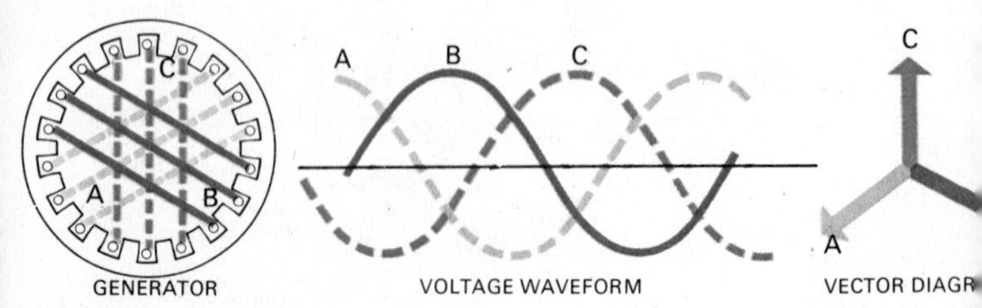

Three phase systems therefore need only half the conductors that three single phase systems require to carry the same power, and are thus much cheaper.

High power transmission lines are controlled so that they always have a balanced load and can thus be three wire systems. Distribution systems serving individual houses are four wire systems to allow for any load unbalance, since each consumer is usually fed from one phase. When many houses are supplied the overall load is surprisingly balanced and the fourth, or neutral, wire can be smaller than the phase wires, although it is usually the same size.

Loads can be connected in two ways. The Delta system (so called because the symbol for the Greek letter delta is a triangle) is a 415V three wire system normally for balanced three phase loads, such as factories using large motors. The Wye system is a 415V three phase or 240V single phase system with a neutral wire for unbalanced loads, such as a combination of three phase industrial and single phase house loads in one area.

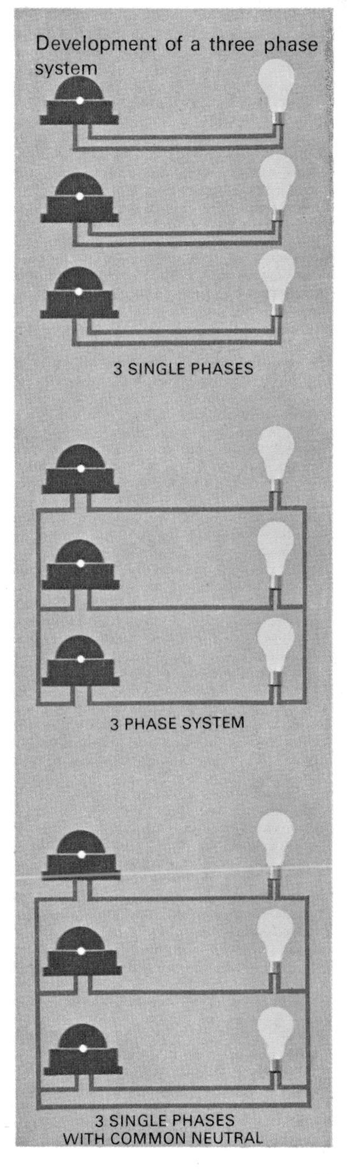

Development of a three phase system

Control of transmission

There are many advantages in having large interconnected power systems. If one generating plant breaks down, its load can be supplied from others in the system, or, if a sudden large supply of electricity is required, it can be drawn from a number of generating stations, instead of overloading one. As loads on a system are lower at night, only a few generating stations are needed at night to supply large areas and these can be the ones with the lowest overall operating costs. The more costly older stations can be used during the day as required. Large generating stations, which are more economical than the equivalent number of smaller ones, are only viable with large systems.

The disadvantage of large systems is the controlling of them. Keeping all the generators working at the same frequency is the main problem, since if they are not, voltage differences occur between power stations and currents flow between them instead of to the load. If a load is suddenly applied to a generator, it slows down, thus reducing the

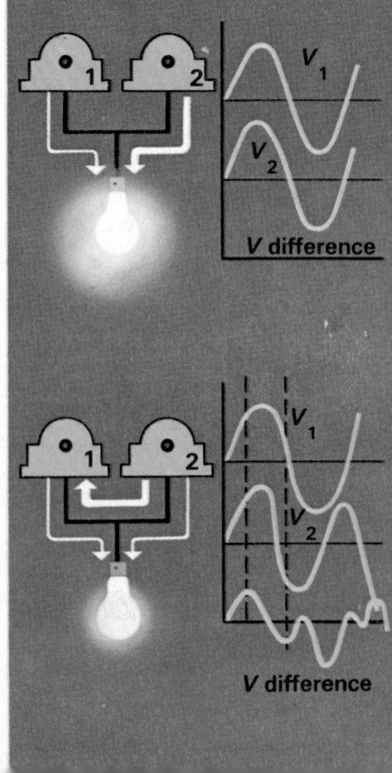

(*Far left*) British grid system
(*Left*) Effect of generator frequency
(*Above*) Grid control room

frequency, and more steam must be supplied to bring it back up to speed. If a load is suddenly removed, the generator speeds up and the steam supply must be reduced. Neither of these operations can be carried out instantaneously.

The British grid system is the largest system in the world under unified control. It consists of 235 generating stations producing 160,000 million kilowatt hours per year with a maximum output of 40,000MW. Electricity is transported by over 8000 miles of transmission lines. The original grid operated at 132kV but this has been improved by the addition of the 275kV supergrid and now sections of this are operating at 400kV.

Control of the overall supply is vested in the National Grid Control Centre in London, which directs the overall power flow. Detailed planning is delegated to eight area control centres. The area centres are in constant contact with each other and the national centre by telephone, telex and radio. These centres plan daily the power flow in advance, according to seasonal trends, the weather, and the day of the week. They then tell each power station how much power to produce and what voltage to generate it at.

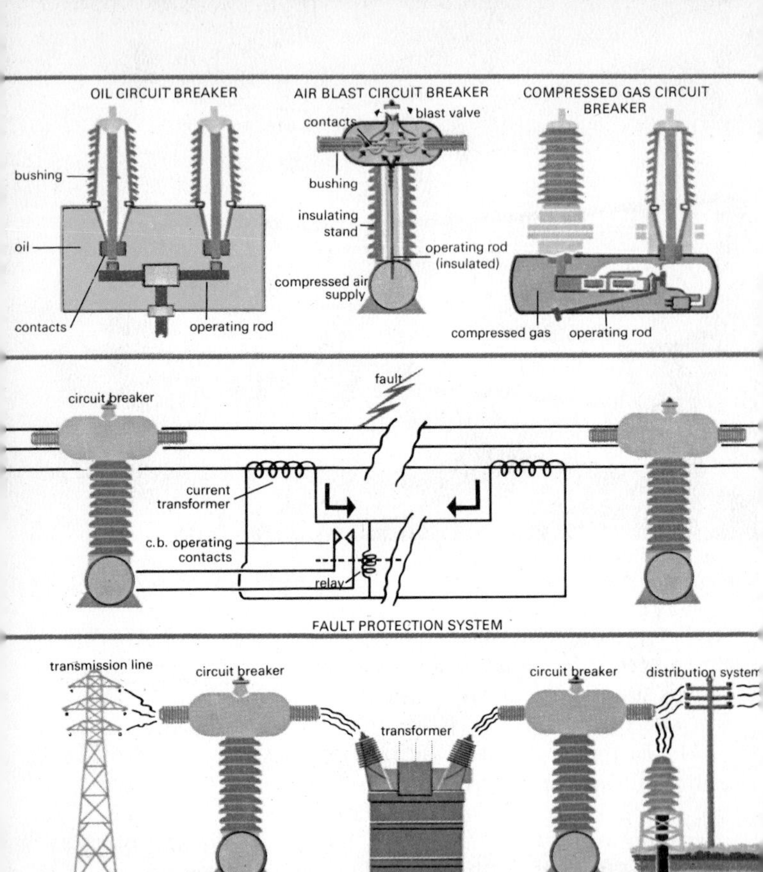

Power transmission systems also have *substations* and *switching stations*. At substations the voltage levels are changed, either up from generating to transmission levels, or down to distribution levels. As well as transformers, substations include circuit breakers, switches for interconnecting the various systems as required, and current and voltage transformers for monitoring the system operation. Switching stations are places where a number of transmission lines of the same voltage come together. There are no transformers, and the lines can be interconnected.

Circuit breakers are necessary to disconnect or connect lines carrying large currents, for an arc forms when normal switches are opened, the heat thus generated melting the contacts. Circuit breakers incorporate special means of extinguishing this arc. Three types of circuit breaker are in general use: the oil-filled type (OCB), the compressed gas type (CGCB) and the air blast type (ACB). In the OCB and CGCB the contacts are in a tank filled with oil or compressed gas which extinguishes the arc. The ACB uses a blast of high pressure air to blow out the arc in the same manner as a candle is extinguished.

Under normal conditions, current interruption is not difficult, but when a short circuit occurs on a transmission line, the circuit breakers are the 'fuses' and must be able to interrupt currents of up to 100 times normal. When faults occur, sensing devices called *protection systems* operate the circuit breakers. In the typical balanced protection system illustrated, as long as the current is the same at 'a' and 'b' the relay is unenergized. If a fault occurs at 'c' the relay operates, opening the circuit breaker and isolating the fault. With this system the circuit breaker automatically closes about one fifth of a second after it has opened and if the fault is clear the line stays in operation. If the fault persists, the circuit breaker reopens and locks itself in this position until it is manually reset.

Numerous other protection systems are in use, some to detect low or high voltage, others to check the frequency, others to sense open circuits, and all indications are transmitted to the area and national control centres.

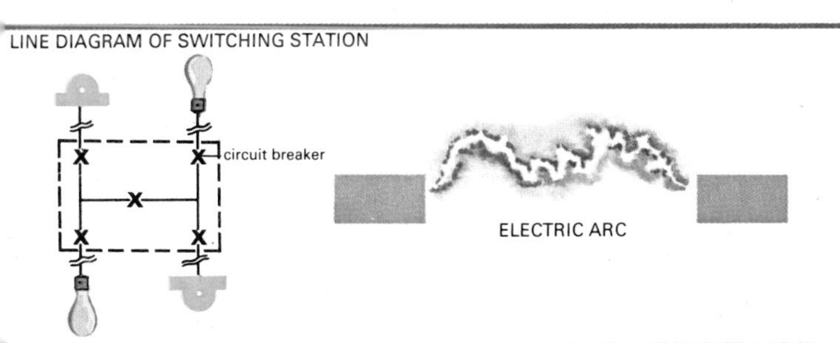

Mechanics of transmission

Power transmission systems comprise both cables and overhead lines. The latter are cheaper but unsightly. They are also vulnerable to lightning strikes. All lines have an earthed wire running along the tops of the towers to protect the phase conductors, but strikes still occur. Cables are used where clearances, for example in city centres or distances across a river, preclude the use of overhead lines.

On many lines, *lightning arresters* are connected from each line to earth. A typical arrester consists of a porcelain housing, containing a number of air gaps in series with a non-linear resistor. The latter is a special material (usually silicon-carbide), the resistance of which decreases as the voltage increases, unlike a normal resistor which is independent of voltage. The flash-over voltage of the gaps is greater than the system operating voltage, but when a lightning surge occurs, they spark-over and conduct the surge to earth through the resistor which has a low value due to the high surge voltage. After the surge has passed (100 microseconds) the resistance increases, which reduces the voltage across the gaps. The magnetic field set up by the heavy current is at the same time stretching the arc. The combination of these effects extinguishes the arc and the device is ready to operate again. This complete sequence happens within half a cycle of the line voltage.

For cables and overhead lines there are length limitations. The critical length for cables is 25 miles on average, because the high capacitance to earth (about 0·2 μF/mile) takes all the available current, leaving none for the load. This can be overcome by *compensation*, in which an inductance is installed between the conductor and earth every 20 miles. The inductance value is such that its reactance is equal to the capacitative reactance of the cable, and as these are 180° out of phase (page 61) they cancel and the cable is effectively lossless. It is not always possible to install compensation where required, for example if the cable is under water, and so d.c. is being considered for long cables to overcome this problem (page 104). With overhead lines the critical length is about 350 miles due to the series inductive reactance increasing with length. All the voltage is required to overcome this impedance leaving none for the load. Series capacitance is added to cancel the series reactance.

Compensating capacitor installation and lightning arrester operation (*right*)

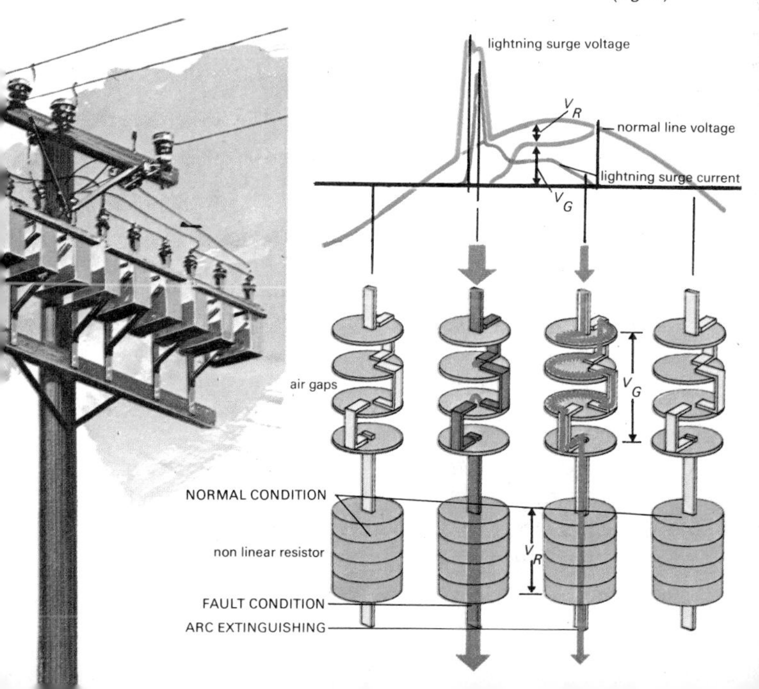

Overhead line systems

An overhead line system consists of three main items: the conductors, the insulators and the towers. Air is the main insulating medium and so it is only necessary to insulate the conductors where they are supported. Copper and aluminium conductors are used. To give adequate strength aluminium lines have a steel core.

On very high voltage, high power lines, *bundled* instead of single conductors are used. These consist of either two conductors in parallel, spaced 12 inches apart, or four in parallel, on a 12 inch square. This is necessary to carry the higher currents, for two or more conductors can carry more a.c. current than a single conductor of the same total cross-sectional area, and to reduce the corona losses, for a four-conductor bundle has a larger *apparent* diameter than a single conductor of equivalent cross-sectional area.

Insulators are usually of the type known as *cap and pin*, but solid rod insulators are used for some low voltage (<33 kV) lines. A number of these porcelain or toughened-glass cap and pin units are connected together with ball and socket joints to support the line, the exact number depending on the voltage involved. This type of construction allows the units to flex when the wind blows, so that they are less likely to fracture. The units are *ribbed* to give a long surface *creepage* path between metal parts, since in service the surface becomes covered with dirt which, when wet by rain, becomes conducting and tends to short out the insulator. The *arcing horns* protect the insulators from lightning: as the air gap between the horns is less than over the insulator, the discharge will be across the horns.

Lines up to 66kV are often supported by wood poles whereas higher voltage ones are invariably supported by metal towers. Steel is the major tower material but aluminium is being used in ever increasing amounts.

Steel cored aluminium (SCA) conductor

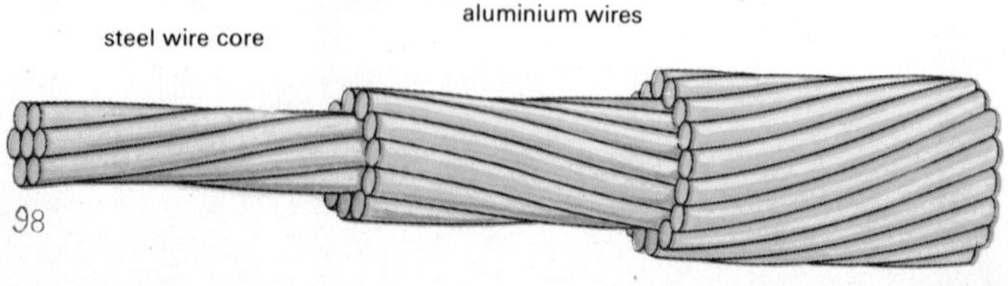

aluminium wires

steel wire core

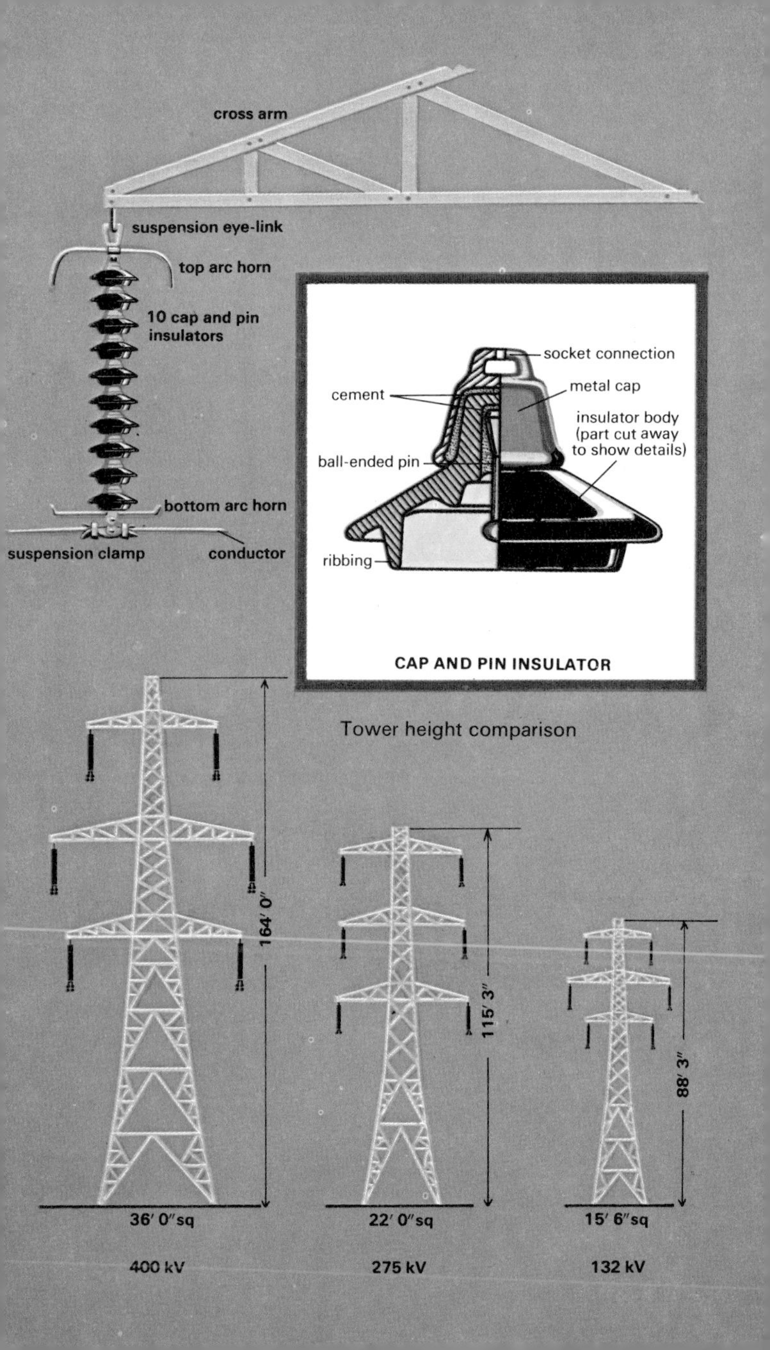
cross arm
suspension eye-link
top arc horn
10 cap and pin insulators
bottom arc horn
suspension clamp
conductor
socket connection
metal cap
cement
insulator body (part cut away to show details)
ball-ended pin
ribbing
CAP AND PIN INSULATOR
Tower height comparison
164′ 0″
115′ 3″
88′ 3″
36′ 0″sq
22′ 0″sq
15′ 6″sq
400 kV
275 kV
132 kV

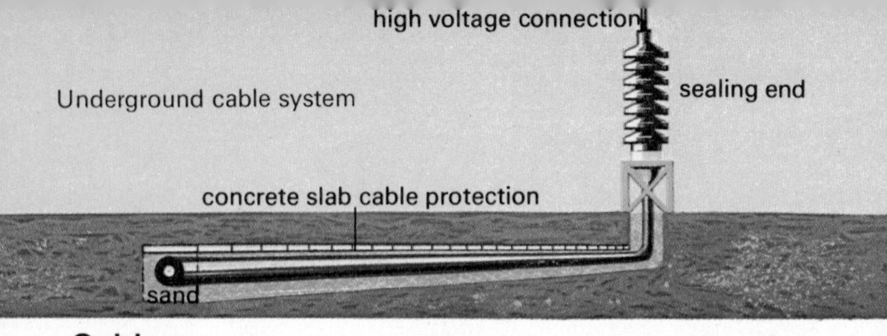

Cables

Cables differ from overhead lines in that the conductor must be insulated along its whole length and protected from damage. Also, to carry the same current as an overhead line system the conductor must have a much smaller resistance (that is, a larger cross-sectional area) as the heat due to the I^2R losses cannot be dissipated as easily underground.

High voltage cables

The structure of present-day cables originated in 1889 when Ferranti laid one between Deptford and London at the then unprecedented voltage of 10kV. It comprised brown paper wrapped around a copper rod, impregnated with wax and inserted into a copper tube. Today cables operate at 400kV and the design of 750kV cables is in progress.

Modern cables consist basically of a conductor (copper or aluminium) surrounded by insulation and protected by a metallic sheath (lead or aluminium) which is in turn protected by an anti-corrosion coating of bitumen or plastic. Up to 132kV, cables usually consist of three cores (for three phase operation) inside a single sheath, while higher voltage cables are of single core construction, mainly for ease of handling. Armour wires (steel) are lapped overall to protect the cable from mechanical damage after it has been buried.

The insulation (or dielectric) of most cables in use today is paper impregnated with oil (oil-filled cable) or an oil/rosin mixture (solid cable). The latter is generally only used up to

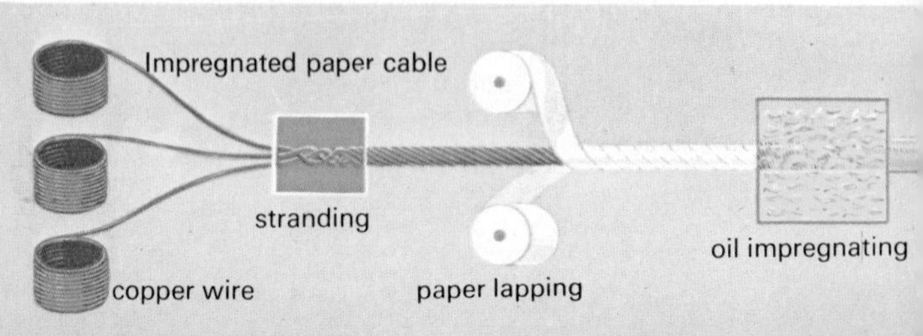

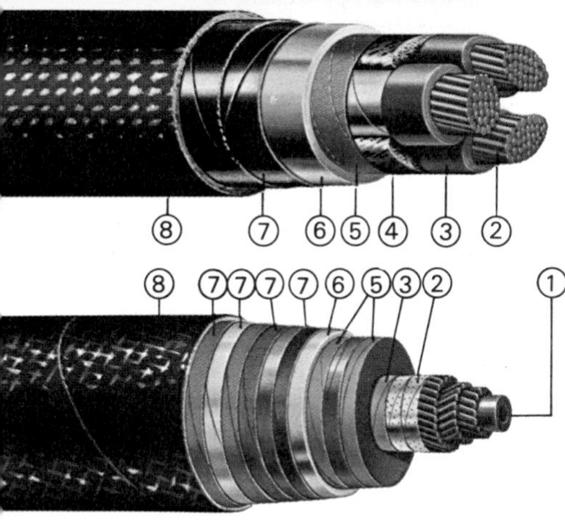

1 Oil duct
2 Conductor
3 Core insulation
4 Fillers
5 Copper-woven fabric tape
6 Lead sheath
7 Anti-corrosion protection
8 Overall serving (hessian and whitewash)

Typical high voltage cables. (*Top*) 3 core 11 kV cable. (*Bottom*) Single core 275 kV cable.

33kV, the oil/rosin being a thick gum. In the oil-filled cable the oil is thin and is maintained under a positive pressure of 75 pounds per square inch which ensures that no airspaces, that can cause premature failure, are present.

During manufacture insulating tape is applied and built up to the required thickness (about 1 inch for 400kV cables) before being placed in a tank, which is then evacuated to remove all traces of air and moisture, and impregnated with oil. The lead or aluminium sheath is then applied, followed by the protective coating and the armour wires.

Oil/paper is being challenged by synthetic insulation, polythene or PVC for example, that is directly extruded on to the conductor (page 103). At present this is limited to voltages below 33kV as the working life of existing materials is short at high stresses. However, a number of experimental cables (up to 138kV) have been installed in various parts of the world and these may be the cable systems of the future.

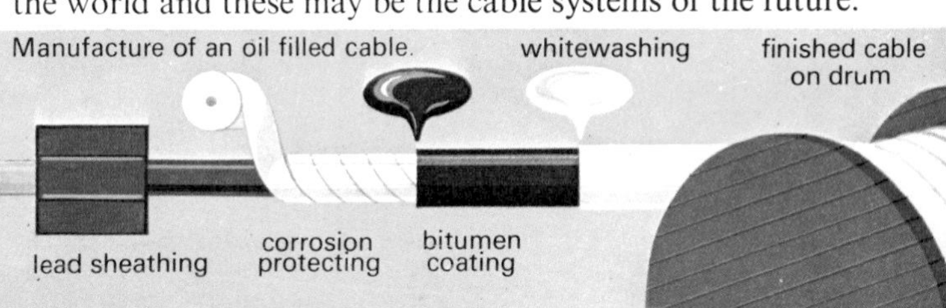

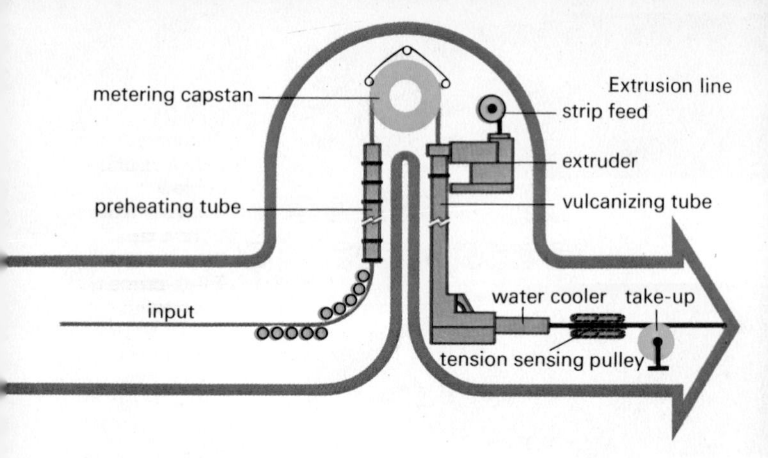

Mains cables

These are low voltage (less than 1000V) cables and were for many years insulated with *varnished paper* or *natural rubber* but are now almost universally insulated with synthetic materials (PVC and butyl rubber). Mains cables consist of an insulated core, or cores, inside a sheath and are armoured for direct burial or unarmoured for interior use.

The conductors (stranded or solid) are sector shaped rather than circular so that a large cross-sectional area can be accommodated in the smallest overall diameter. Aluminium has supplanted copper as the major conductor material for these cables.

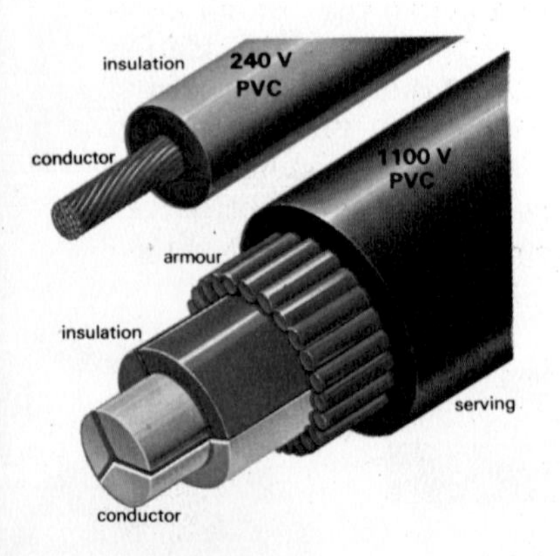

(*Left*) Mains cables. (*Right*) Cable under stringent testing.

Plastic cables are manufactured by *direct extrusion* of the insulating material on to the conductor. The material is melted, applied to the wire through dies to give the required thickness and then cooled in water troughs. This operation is carried out at speeds up to 1000 feet per minute.

Cable cooling

The major problem in buried cable systems is dissipating the large quantities of heat generated by the conductor I^2R losses. The heat must pass through the insulation and into the ground to escape. Thus two or more cables of much larger area than the overhead line conductor are needed to carry the same current.

A major portion of cable research at the present time is directed towards solving the cooling problem so that fewer (to keep trenching costs down as this is a major cost in installing a cable system) and smaller cables can be used.

A current method involves circulating water in pipes buried alongside the cables. This has allowed a 50% decrease in conductor size on two cable routes in London. Cooling efficiency is still low as the heat must still come out through the insulation to get to the water pipes. Another method under development is to circulate oil through a duct in the centre of the conductor, thus removing the heat at source.

Oil-cooled cable

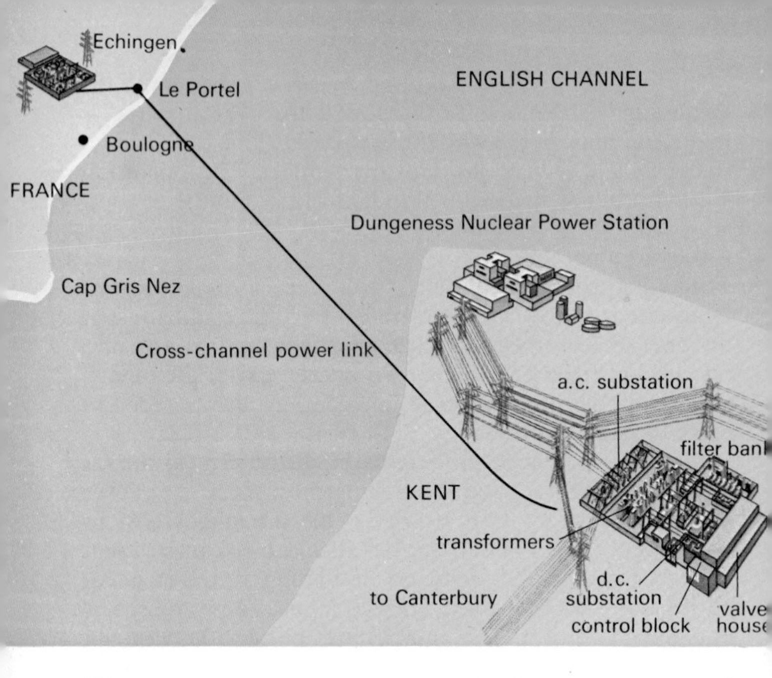

Direct current power transmission

The alternative to high voltage alternating current (HVAC) power transmission is high voltage direct current (HVDC). The earliest systems (before 1900) were mainly d.c., but with the advent of the transformer for stepping up a.c. voltage, d.c. was discarded. However, modern developments in conversion equipment have made the changing of a.c. to d.c. and back again possible at voltages up to 450kV.

A d.c. system has many advantages over an a.c. one. It has no inductance and capacitance effects and thus no stability problems or cable length limitations. It will transmit more power for less copper and operate at higher voltages for the same amount of insulation and is, therefore, cheaper and more compact. However, the rectification equipment is complex and expensive and circuit breakers are not available. Thus the overall cost is only lower for d.c. systems of very long lengths.

Large controlled mercury arc rectifiers are used in all existing and proposed schemes. The amount of power flowing is governed by the current flow which is controlled by the portion of each a.c. cycle that the rectifiers are conducting. Electronic control enables faults in d.c. systems to be isolated more rapidly than in a.c. ones so that less damage is caused. High voltage thyristors (page 49) are being developed for these systems.

The main uses of d.c. transmission systems are at present in power transmission over long distances (New Zealand, Russia and the proposed United States West Coast System), and in long submarine cables (Sweden and Vancouver Island). In Japan, where half the country works on 50 hertz and the other on 60 hertz, a d.c. link has made it possible to interconnect the two systems. The England-France scheme allows each country to reduce the amount of generating stations it must build to cater for peak demand, because this occurs at different times in the two countries and power can be moved either way as required, without having to keep the two large a.c. systems in perfect synchronization, which would be very difficult.

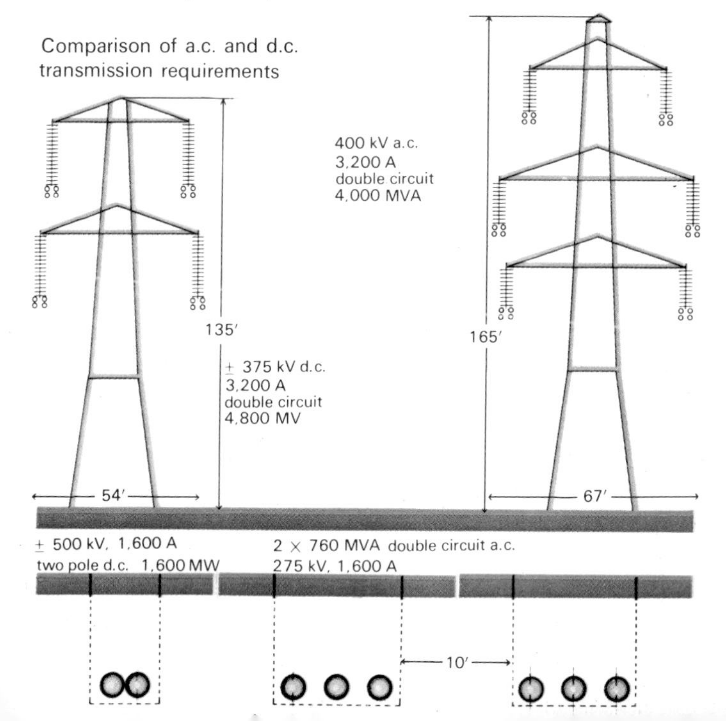

Comparison of a.c. and d.c. transmission requirements

ELECTRICITY IN INDUSTRY

Electricity is the life blood of modern industry and our standard of living would be much lower without it. This section shows how the four basic effects of electricity, heat, magnetism, chemical action, and charge storage, either singly or in combination, are utilized in industry.

Increasingly these processes are becoming more complex so that human and mechanical control are now far too slow and unwieldy for many operations. Electronic and computer control is briefly discussed but fuller details are available in *Electronics* and *Computers at Work*.

If electricity is the life blood, then electric motors are the arms and legs of industry. Without them factories would still be powered by one large steam engine driving all the machines by an overhead belt system. Individually mounted electric motors have made modern factories safer and more flexible, efficient and pleasant.

Electric motors

As shown earlier, a current is induced in a wire when it cuts a magnetic field and if a current flows in a wire, suspended in a magnetic field, the wire moves, due to interaction of the applied magnetic field and the induced one around the wire. This is the basis of electric motor operation.

Motors are designed to operate from a.c. or d.c. supplies or both. The electrically produced magnetic fields cause the moving part, the *rotor* or *armature*, to revolve inside a fixed frame, the *stator*.

Constructionally, motors are virtually identical to generators. Basically, they consist of a circular steel frame (stator) with the windings wound around steel poles attached to the inside. The rotor is supported on bearings at each end, with the windings set into longitudinal slots on the circumference of the cylindrical steel core. Commutators and slip rings are copper and the brushes are a mixture of carbon and graphite. The air gap between the rotor and the stator must be kept as small as possible, since air has a high resistance to magnetic flux.

Electric motors vary in size between the domestic, fractional horsepower sizes, and the industrial, 10,000 horsepower motors. One horsepower is about 0·75 kilowatts.

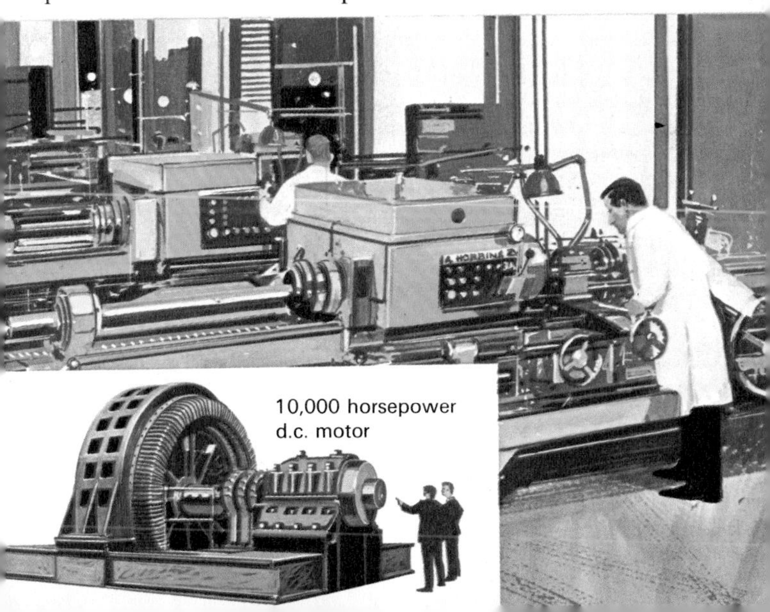

10,000 horsepower d.c. motor

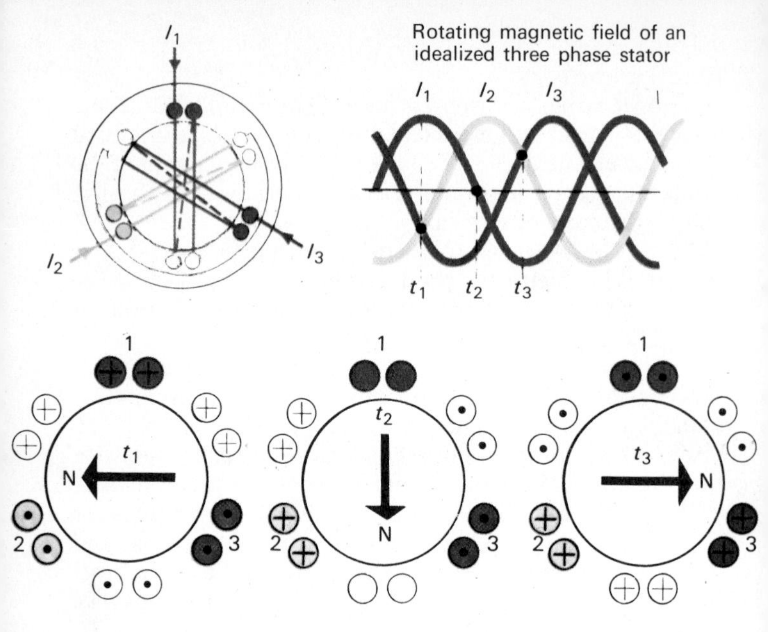

Alternating current motors

In three phase machines an effectively rotating magnetic field is set up in the stator as follows. The magnetic field changes with time and is illustrated for three, marked instants in time. Firstly, at t_1, I_1 is positive (above the axis) and I_2 and I_3 are negative. Thus the currents are flowing in the directions indicated (+ is into the page, · is out of the page) and by using the *right hand rule* the magnetic field is as shown by the arrow. Then at t_2, I_1 is zero, I_2 is positive and I_3 negative and the magnetic field has shifted. Finally at t_3, I_1 is negative and I_2 and I_3 are positive and the field has again shifted. Thus, as the currents are cyclically repeating the magnetic field rotates, and if the rotor, which is inside the stator, is a magnet, it will be turned.

This is the principle of the *synchronous motor*, so called because it runs at a constant speed that depends on the a.c. supply frequency (page 44). This means that they cannot start themselves, since the stator field is moving too fast for the rotor to catch up. In very small motors a permanent magnet rotor is used but in all others a d.c. energized electromagnet is used, the current being fed in via slip rings. Their main uses are in driving large loads, where constant stopping and starting is not required, such as d.c. generators, pumps and compressors.

Induction motors (the commonest used in industry) have windings on the rotor identical to those on the stator. By transformer action, a rotating magnetic field is induced into the rotor windings, causing the rotor to rotate. An induction motor always runs slightly below the rotational speed of the magnetic field, otherwise no magnetic lines of force would be cut and thus no voltage induced in it. At standstill, the maximum current is being induced in the rotor and thus a high torque is produced so that it is a self starting machine.

The rotors of synchronous motors are usually wound as for induction motors. They are then self-starting, running up almost to synchronous speed before the d.c. is applied to the rotor, when they operate normally.

With single phase supplies there is no rotating field and so a motor will not start, but once in motion it will keep running. By having two windings, one in series with a capacitor, a quasi-rotating field is set up due to the phase shift of the current through the capacitor. A centrifugal switch removes the capacitor from the circuit when operating speed is reached.

Induction motor

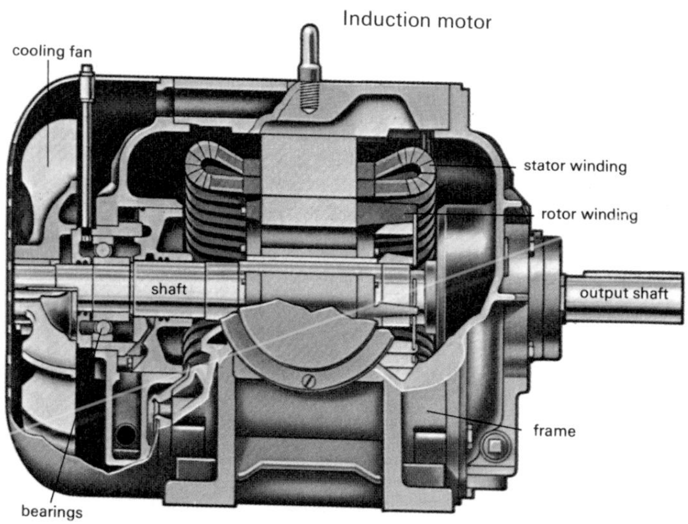

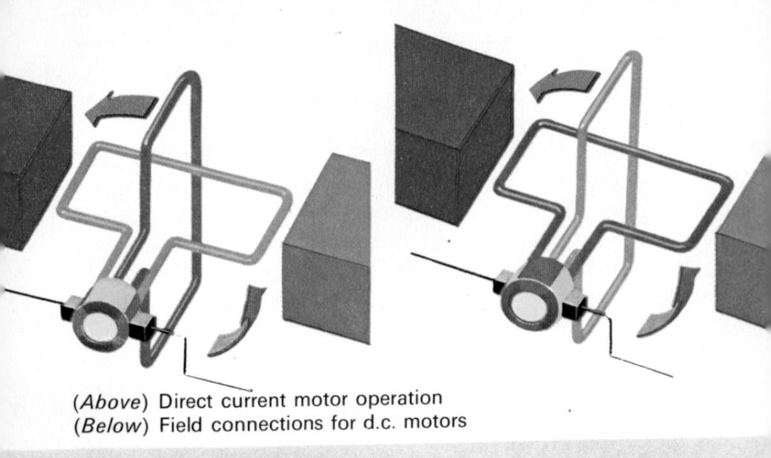

(*Above*) Direct current motor operation
(*Below*) Field connections for d.c. motors

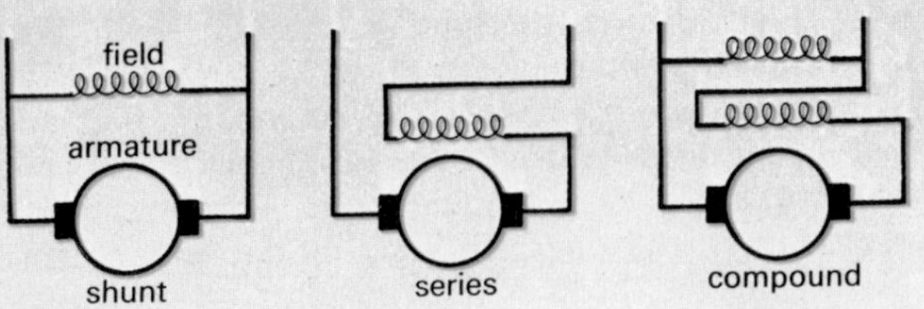

Direct current motors

The principle of d.c. motor operation is slightly different from that of a.c. being based on a *partially rotating field concept*. These motors have a stationary magnetic field in the stator and a number of coils on the armature in which magnetic fields are set up in turn from current fed via the commutator. Initially the armature coil field is as shown by the arrows in the diagram and the coil tends to rotate to line up with the stator field but, as it turns, the commutator switches the current to the next coil and so continuous rotation results. Thus d.c. motors are also self starting.

There are three basic ways of interconnecting the field (stator) and armature windings, each one giving a different characteristic. Series motors have high starting torques but their speed drops by 60% between light and full load so that they find their main application where heavy loads must be repeatedly started and stopped. The shunt motor has a low starting torque but the speed only drops by 3% and is used where constant speed over a wide range of loads is required. Compound motors are midway between the other two in starting torque but only drop their speed by 15%.

Universal motors

The series d.c. motor will also operate on a.c. because the same current flows through the armature and the field, so that both fields change polarity at the same time. Thus, when the field polarities reverse, the absolute polarities have changed (N to S and S to N) but the relative polarities are still the same (opposite) and therefore the rotation continues. When the brushes bridge two commutator segments heavy currents are set up by transformer action, causing a large spark. This limits the size to under one horsepower but, because of its versatility, this motor is used in many home appliances and power tools.

Speed control of motors

Some applications require motors whose speed can be varied over a wide range. The only way to obtain this with a.c. motors is to vary the frequency but this is an expensive and complicated undertaking. Limited speed control of induction motors is possible, by reducing the voltage or adding resistance into the rotor circuit, but this reduces the efficiency markedly and the motors tend to stall easily. The speed of a d.c. motor can be varied from virtually zero to normal speed simply by placing a variable resistance in series with the armature, to limit its current. For this reason, d.c. motors have never been supplanted by a.c. motors.

Direct current motor

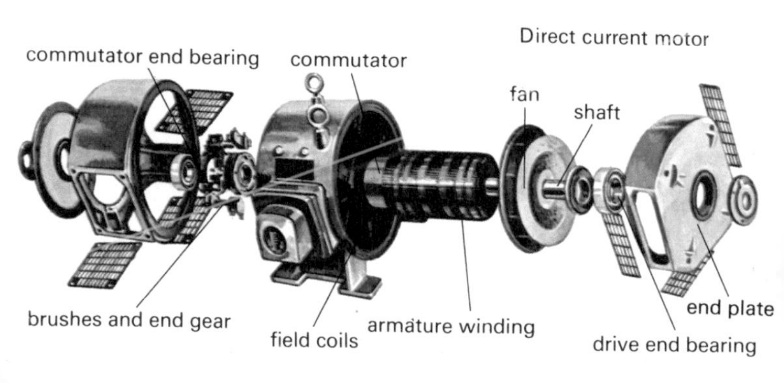

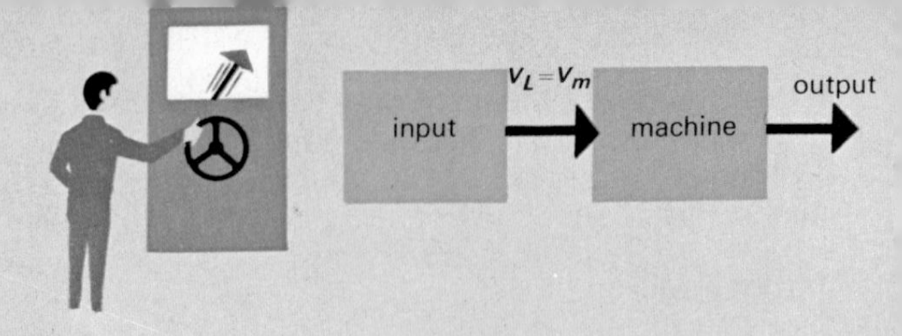

(*Above*) No feedback. (*Below*) With feedback

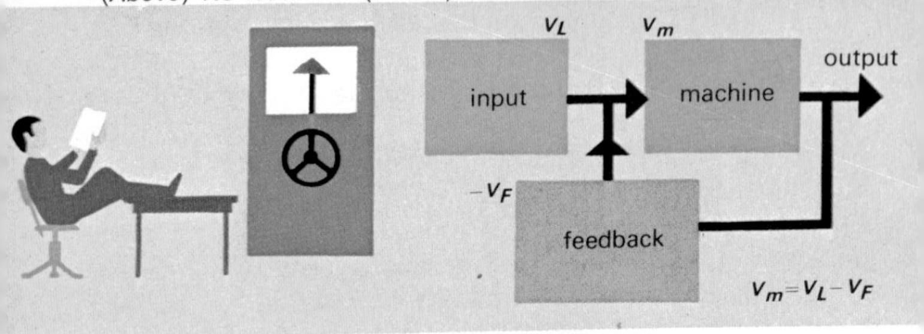

Automatic control

As machines become larger and faster, and especially where a number of them are connected in an interrelated system, it becomes more and more difficult for a human operator to control them accurately. Inevitably, a time lag occurs between something going wrong and it being put right. Automatic control of machines, motors, and systems is thus becoming vital to industry.

Basically, all automatic control depends on *feedback*, where the output of a machine (for example the speed of a motor) is monitored and if it deviates from the normal, a signal is fed back to the input to correct it. In a non-feedback system, the operator tries to keep the operation constant, whereas with feedback he just keeps an eye on it to make sure that the system is operating correctly. This saves manpower, since one man controls many machines.

Negative feedback is the usual mode, that is, the signal from the output opposes the input to the machine. Thus, if the output increases, the feedback increases and as this opposes the input, the input drops and thus the output drops. The process works the other way if the output falls. If the feedback system shown is a motor speed control system, this feedback unit is a small generator connected to the motor output, which gives a signal (V_F) proportional to the speed of the main motor. If the motor speeds up, V_F increases and the motor supply voltage (V_m) is reduced so that the motor slows down to its original speed. If the supply voltage (V_L) is kept constant the motor will always run at the same speed, no matter how much the load varies.

This type of system is used where a large number of machines must run at the same speed, for example in newspaper production. A generator's output voltage can be controlled in a similar manner. A portion of the output is fed back to the field windings so that if the output increases the field decreases and brings the output back to normal.

Increasingly, electronic sensing devices and computers are being employed to control hundreds of machines (see the companion volume, *Computers at Work*). An example of a sophisticated computer controlled system is the detection and interception of enemy aircraft or missiles.

Computer controlled feedback system

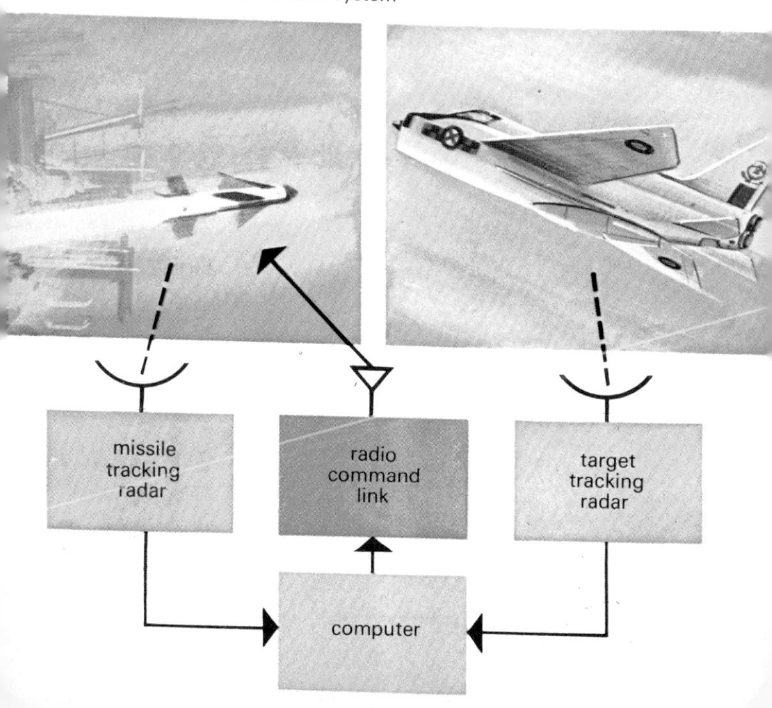

direction of motion

hot area

current carrying coil

cooling water spray

PRINCIPLE OF INDUCTION HEATING

cold area

Electric arc welding is simpler and safer than the usual 'burning gas' processes and the equipment is compact and portable. The arc gives off a brilliant light of mainly ultra-violet rays, which are harmful to the eyes and skin, and so protective clothes and goggles are worn.

Electrical Heating Apparatus

Welding

The basic principle of electric arc welding is that an arc is struck between the workpiece and a consumable *welding rod*, the material of which depends on the metals being welded. The heat of the arc vapourizes the welding rod on to the hot workpiece, thus joining the two parts. The workpiece is connected to one side of a low voltage (35 to 70V) high current (50 to 500A) a.c. or d.c. supply and the welding rod is connected to the other. The rod is then made to touch the workpiece, withdrawn slightly to form the arc, and then moved over the area required to be welded.

A slight variation is MIG (Metal Inert Gas) welding, which is used mainly for aluminium. A hard surface oxide layer forms when aluminium is heated in air (due to reaction with oxygen) and this prevents the metal being joined. This is overcome by surrounding the working area with an inert gas (usually argon) which does not react with aluminium. It is supplied from a nozzle incorporated in the welding rod holder.

Arc furnace

Heat liberated by an electric arc is used in steel mills to melt the metal. Furnaces are filled with cold steel and connected to one side of the supply, the other side being connected to carbon electrodes which are slowly lowered until an arc is struck. Both the heat of the arc and the heat generated by the heavy currents that flow heat the metal.

Induction heating

Electric currents, producing heat, are induced in a conductor when it is placed inside a magnetic field. In practice, the sample is placed inside a coil which is carrying a heavy, high frequency current. The coil is kept cool by circulating water through the hollow conductors. This method is suitable for continuous processes, such as annealing wire, since no physical contact is required and the material can be drawn through the coil continuously. Because the magnetic coupling is through air, high frequencies must be used to attain high efficiencies.

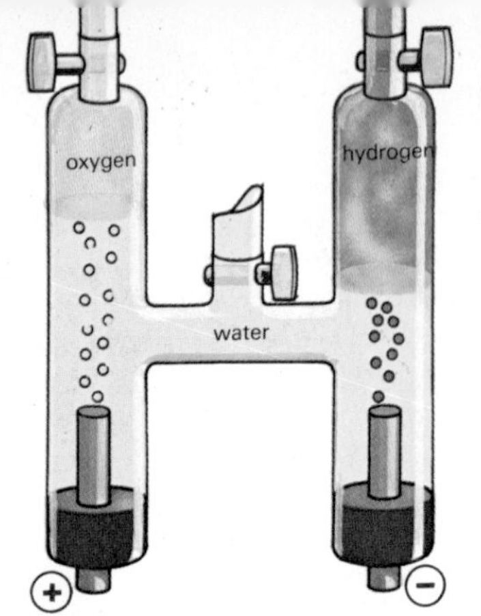

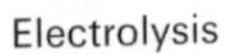
Electrolysis

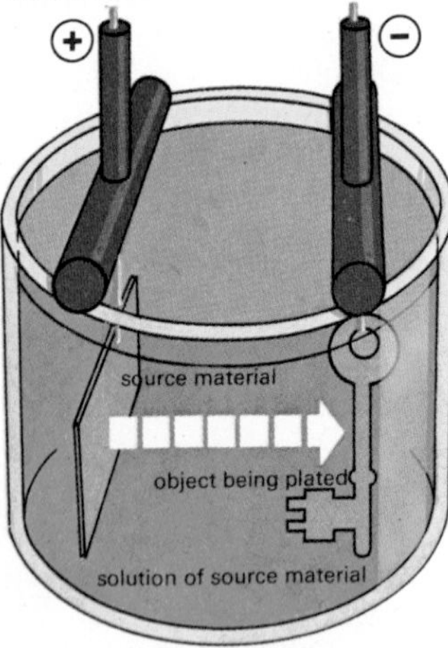

(*Above*) Electroplating
(*Below*) Aluminium smelting

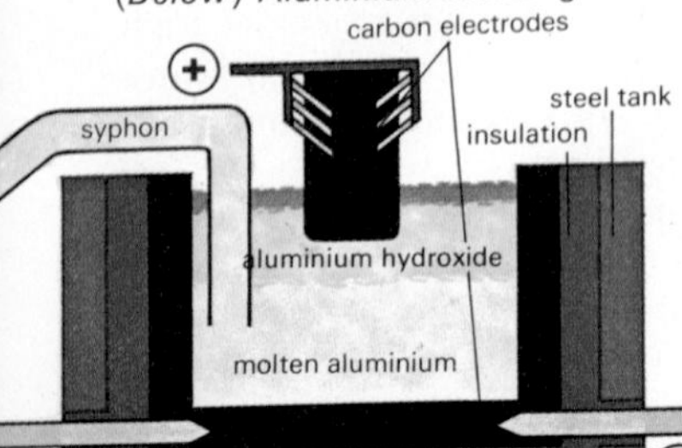

Electrochemistry

When a direct current is passed through a liquid a chemical phenomenon, known as *electrolysis*, occurs, resulting in the liquid separating into its constituent parts. For example with water, hydrogen gas (having a positive charge) appears at the cathode and oxygen gas (negative charge) at the anode. Twice as much hydrogen is produced, since water (H_2O) is composed of two parts hydrogen to one of oxygen. If the gases are removed and the water continually replenished, a continuous production process for hydrogen and oxygen is available. A very pure form of almost any element can be prepared in this manner.

An allied phenomenon is *electroplating*. When two electrodes are suspended in a solution of the anode material and a heavy direct current is allowed to flow, the anode material is deposited on to the cathode, the anode being eroded in the process. Any metal may be deposited on any other. *Galvanizing*, where a rust preventive zinc coating is deposited on iron components, is carried out in this manner using currents of 200 amperes at 5 to 10 volts. The deposited metal is very pure, any impurities (which are normally non-metallic) remaining on the anode or in the solutions. Using large vats, high purity copper, for electrical uses, is produced on a commercial scale. A small rod of pure copper is used as the cathode and a block of impure copper as the anode.

If aluminium is placed in a bath of sulphuric acid and about 3 amperes at 50 volts d.c. applied, the surface of the anode oxidizes and becomes very hard. This *anodizing* process makes aluminium very wear resistant, and also gives a *non-stick* surface that is used for pots and pans.

Large amounts of electricity are consumed in the production of aluminium. The ore (bauxite) is roughly purified into aluminium hydroxide which is then placed in large *electrolytic cells.* Fifty to seventy thousand amperes are passed through the cell using carbon electrodes, the temperature rising to 1000°C. Molten high purity aluminium settles at the bottom of the cell, to be removed by large syphons.

Electrodialysis is a method of separating the components of ionic solutions and is most useful for *desalination*, the purification of salt water. Saline water is continuously introduced into a tank where the electrodes are separated by two semi-permeable membranes. One membrane passes only positive ions, the other only negative, so that, under the influence of the electric potential, the negative chlorine and positive sodium ions pass through the respective membranes, leaving pure water in the central compartment and brine in the outer ones.

Electrodialysis

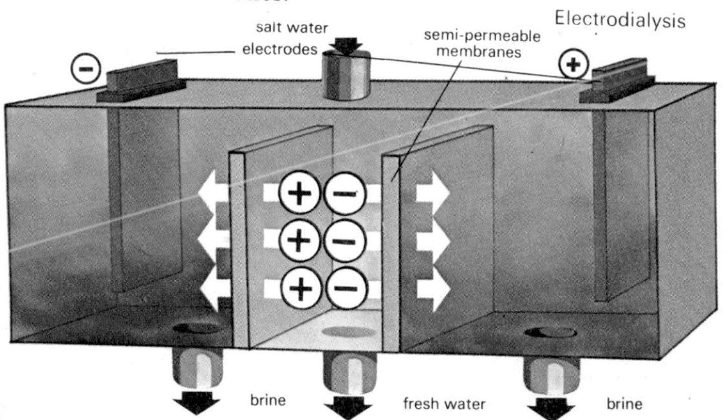

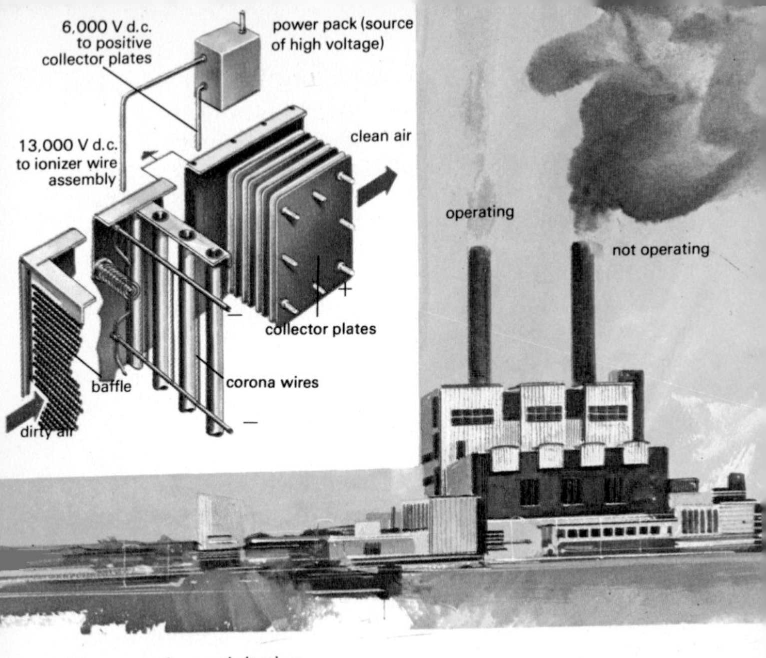

Electrostatic precipitation

Electrostatics

The phenomenon of electrostatics can be dangerous, in hospitals or spacecraft for example, but in industry its applications are increasing rapidly.

The most common use is in *electrostatic precipitators* that remove contaminants from furnace exhaust gases. These gases flow through a row of wires, alternately energized at high negative voltage and earthed. They then flow through a row of plates, alternate ones either being energized at a positive voltage or being earthed, before being released to the atmosphere. The corona discharge from the wires gives a negative charge to the dirt particles in the gas and these are then attracted to, and adhere to the positive plates, which therefore have to be cleaned periodically. Precipitators in modern 2000MW power stations are 99·3% efficient, removing 100 tons of dirt from 330 million cubic feet of exhaust gas per hour.

Spray painting techniques have been advanced by the

use of electrostatics. It is now possible to paint both sides of a workpiece and awkward corners, from one side only. By charging the spray gun and the workpiece at opposite potentials, the paint is attracted to all surfaces. This also reduces wastage, since all the paint is attracted to the workpiece and is spread more evenly. A variation of this method is used in dip painting, of car bodies for example. The body shells are charged and then dipped into a bath of paint which is oppositely charged.

Tea leaves can be separated from the stalks and fibres very easily using electrostatics. After drying, the leaf contains less water than the waste. The moister impurities are attracted by an electric field while the leaves drop under the influence of gravity, undeflected by the electric field. Voltages of 5 to 20kV are used and 2000 pounds of tea per hour can be separated.

The Xerox duplicating process relies on electrostatic charges to form the patterns of the letters to be reproduced. The image of the document to be reproduced is reflected on to a plate of charged, light sensitive material, usually selenium. This plate is discharged in proportion to the amount of light falling on it from the document, the letter patterns thus remaining charged. Powdered ink, of opposite charge, is applied to the plate and adheres to the charged areas only. Copies are printed on to blank paper from this inked plate, the transferred ink being finally fused to form a permanent record by heating. The plate is then wiped clean, recharged and used again. In an actual machine this complete process is carried out automatically and copies can be produced at a rate of two a second.

Electrostatic spray painting

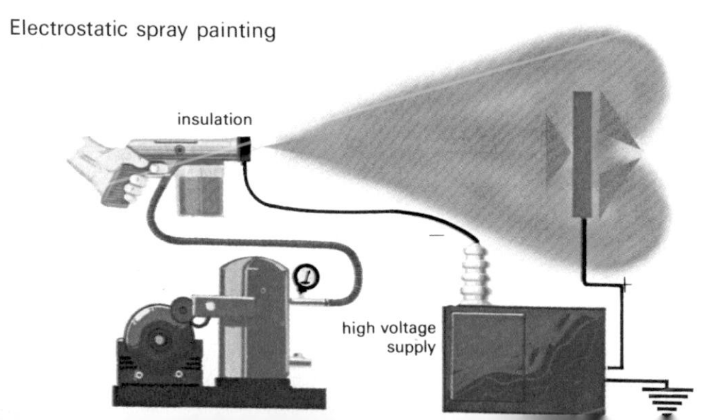

Electron microscopy

Optical microscopes are limited to magnification of less than 2000, because details smaller than the wavelength of the light used cannot be seen. With visible light, details less than 2×10^{-5} inches cannot be observed. Electrons have a much smaller wavelength than visible light so that when it was discovered in the 1920s that a magnetic field could be used to focus electron beams, in a manner analogous to a glass lens focusing a light beam, the electron microscope was born. Magnification up to 100,000 is possible, which is approaching that required to view the structure of matter (page 20). At present, individual atoms have not been observed, but groups of ten carbon atoms have.

Two types of instrument have been developed, the Transmission Electron Microscope (TEM) which looks through the sample and the Scanning Electron Miscroscope (SEM) which displays the sample surface like a binocular microscope. Both these microscopes are illustrated.

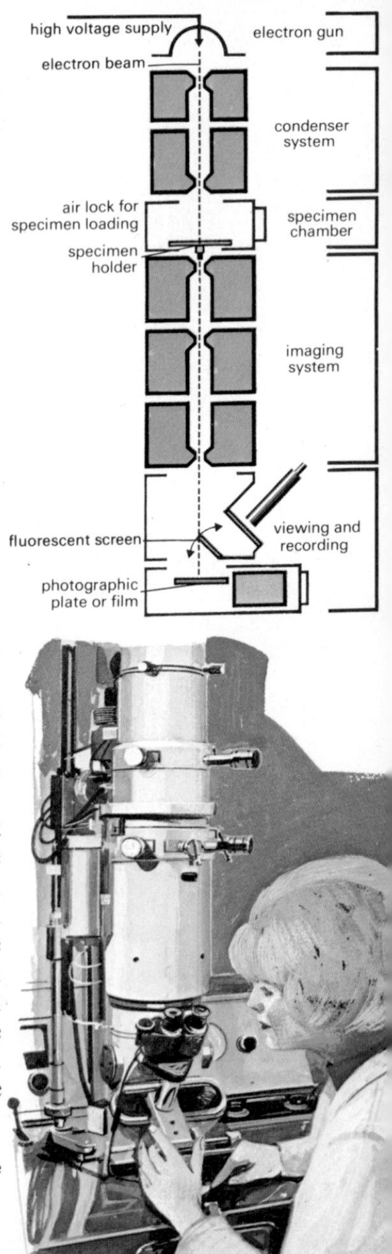

Transmission electron microscope

Basically, both types consist of an evacuated column (otherwise the electrons would collide with molecules in the air and be dispersed) with an *electron gun*, a source of electrons, at the top. This gun is at a high negative d.c. potential, usually supplied by a van de Graaf generator, and the electrons are thus accelerated towards the earthed end of the column, being focused into an extremely fine beam by the magnetic lens system.

In the TEM, the electron beam passes through the sample, which must be very thin or the electrons will be completely absorbed, and then on to a fluorescent screen for observation or a photographic plate for a permanent record. This requirement of a thin specimen limits the resolution to date, since samples less than one hundredth of a millimetre thick cannot be prepared. Thus, only bulk, and not surface details can be examined.

To overcome this last disadvantage the SEM was developed. When the beamed electrons strike the surface of the specimen, the majority enter but some bounce off and are collected and used to control the brightness of a cathode ray tube. The number produced depends on the angle and distance the surface is from the electron gun. The electron beam is *scanned* across the specimen by means of deflection coils and a picture is built up, in the same manner as on a television screen, of the surface of the object. This display can be observed and photographed. Magnification is only 20,000 but a three dimensional image is produced, thus giving a much clearer picture than an ordinary microscope.

Scanning electron microscope

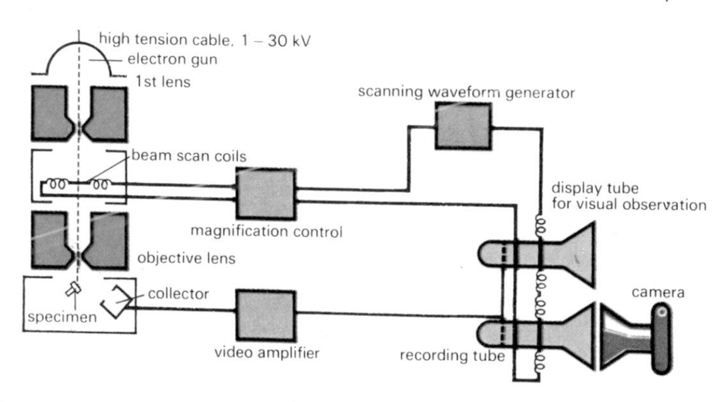

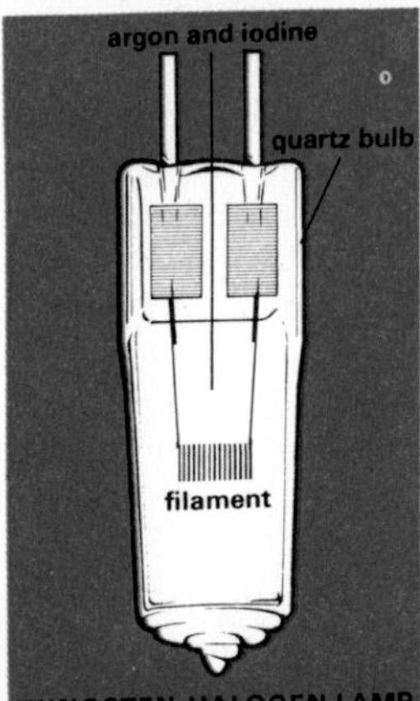

TUNGSTEN-HALOGEN LAMP

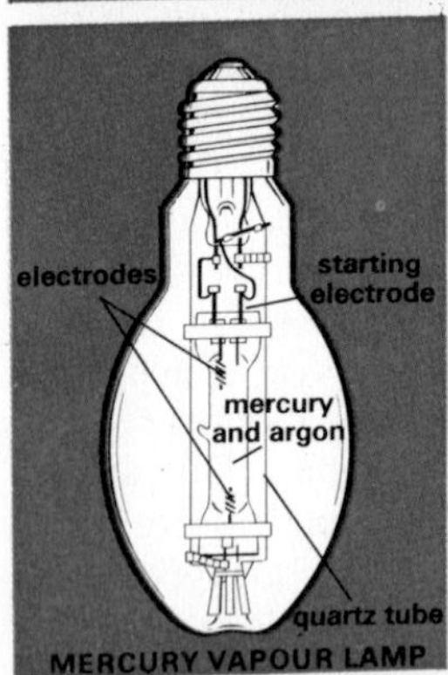

MERCURY VAPOUR LAMP

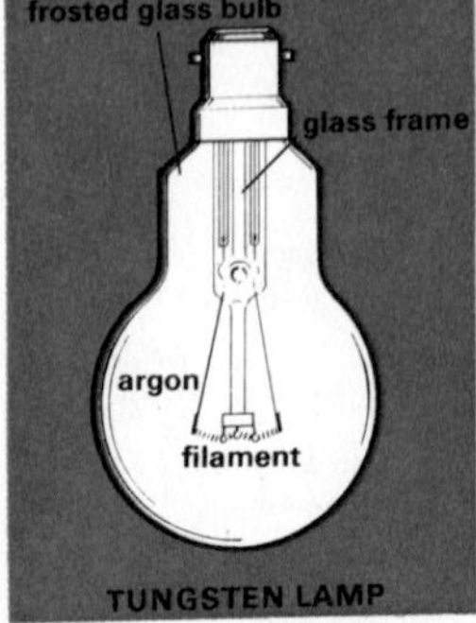

TUNGSTEN LAMP

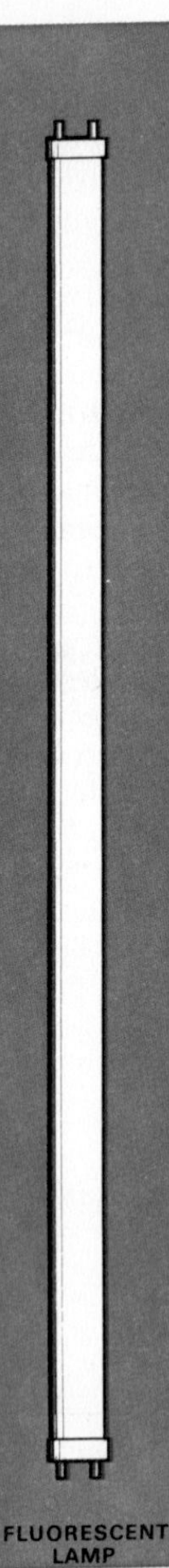

FLUORESCENT LAMP

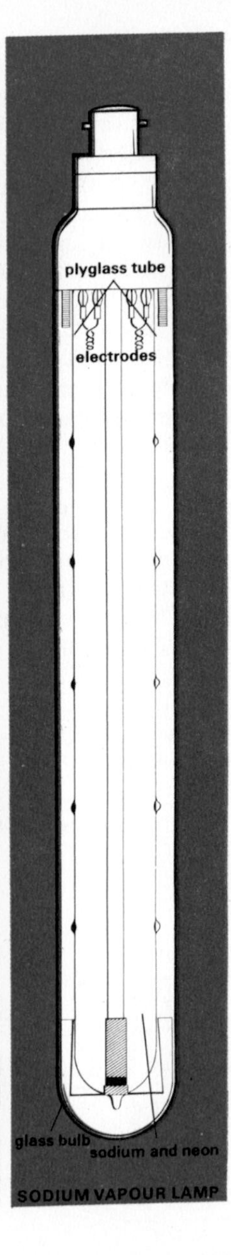

SODIUM VAPOUR LAMP

Illumination

The tungsten or incandescent lamp is the commonest in use. The light source is a thin filament of tungsten wire, raised to white heat by the current flowing through it. It is contained in a glass bulb, filled with inert gas (usually argon) at low pressure, to prevent the filament burning away and the glass bulb getting too hot. These lamps are very inefficient and have a short life, since the filament gradually evaporates. Krypton gas filling promises an increased life span.

To overcome this filament evaporation, traces of iodine vapour are added to the argon, causing the evaporated tungsten to be redeposited on to the filament instead of the bulb. This necessitates a higher bulb surface temperature and so quartz is used instead of glass. These tungsten-halogen lamps are more efficient and have twice the life of ordinary incandescent lamps, but are much more expensive. They are used in projectors, cars and flood lights.

Sodium and mercury lamps work on the principle that a large current flowing through a gas heats it, causing it to emit visible *radiation*. Sodium lamps contain sodium and neon gas at very low pressures. When the operating temperature is reached the orange glow of sodium predominates. The human eye is most sensitive to orange and so these lamps provide the best illumination for night vision. The mercury lamp uses a mixture of argon and mercury gas pressurized at one atmosphere. Its output is mainly blue but it provides a greater output for a physically smaller size than sodium lamps. Both types are efficient.

Fluorescent lamps are similar to the mercury vapour lamp but have the inner face of the glass tube coated with phosphorescent material (calcium or zinc) and use low pressure mercury vapour. The radiation is mainly invisible ultra-violet, but this causes the coating to *fluoresce* in the visible light region. These lamps have twice the efficiency of incandescent lamps and operate at a lower temperature.

A similar type is the neon lamp used for illuminated signs. A high voltage is applied between the ends of the neon filled glass tube. A low current glow discharge forms in the gas, red being the characteristic colour of neon but other colours being obtained by adding traces of other gases.

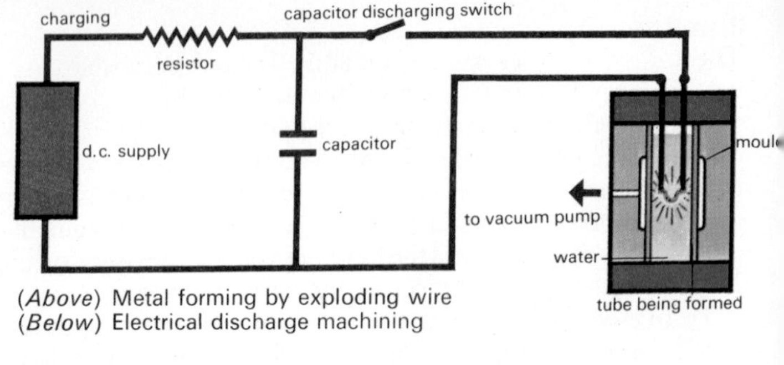

(*Above*) Metal forming by exploding wire
(*Below*) Electrical discharge machining

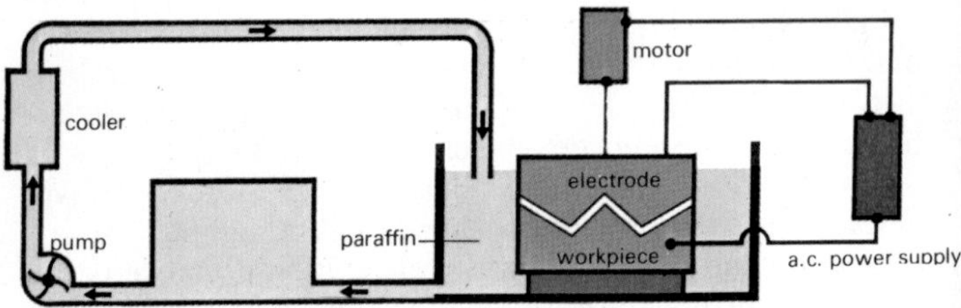

Metallurgical uses

Metal forming

When an electric current flows for a very short time, a considerable amount of energy is released. This energy can be utilized to easily form metal sheets into shapes that would otherwise be virtually impossible to create. In the illustration a metal tube is being shaped. The sudden discharge of the charged capacitor forms an electric arc, the sudden increase in heat causing a shock wave which travels through the water (water is a better pressure transmission medium than air) until it hits the metal, causing it to bulge into the mould. The mould is evacuated to reduce the resistance to the movement of the material.

Electrical discharge machining

In this contactless method, the workpiece is shaped to the form of the electrode by high frequency discharges occurring between them as the electrode is slowly lowered on to the workpiece. Local thermal action, produced by the discharges, erodes the workpiece, which must be softer than the electrode. This takes place in a bath of a liquid dielectric, usually paraffin, to cool the workpiece and increase the voltage at which discharge commences (the energy involved is proportional to the square of the discharge voltage). As the frequency of the applied voltage increases, less material is removed, giving a better surface finish.

Plasma arc machining

A high temperature plasma is generated by discharging a powerful electric arc through a gas mixture (nitrogen/hydrogen or argon/hydrogen). The temperature of the gas is raised to a point where the bonds between the molecules break, releasing energy in the form of heat. The main use of this is as a cutting torch for very hard materials.

Electron beam machining

A beam of high energy electrons is focused to a very fine spot on a surface, producing local heating and consequent erosion. Only small workpieces can be handled since the energy dissipated is low. Small objects can also be welded and, as the beam can be accurately positioned, this method is used to attach leads to transistors and micro-circuits.

(*Left*) Electron beam welding. (*Right*) Plasma torch

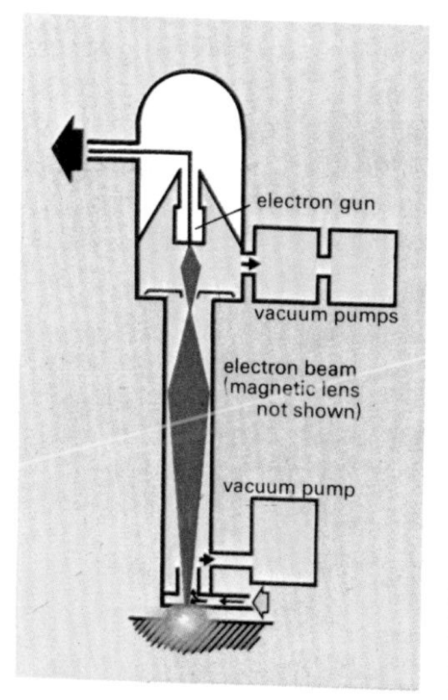

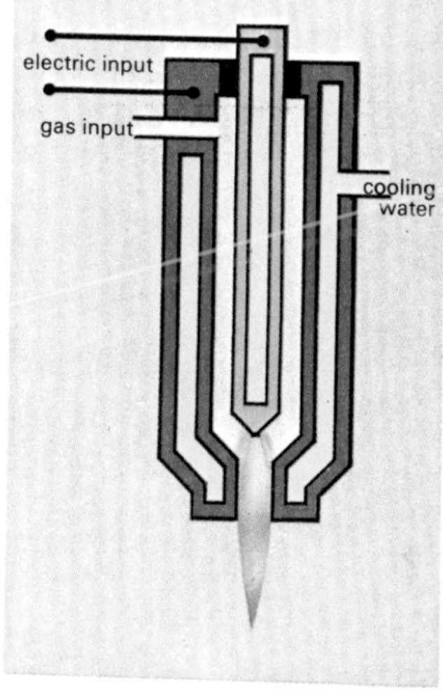

Non-destructive testing

For many years the only method of testing materials, such as steel beams, was by selecting a few at random from a production batch and testing them to destruction to ensure that the failure strength was well in excess of the required service value. Recently non-destructive testing (NDT) techniques have arisen and electricity has contributed to their success. Internal flaws are detected without physical testing and so all units in a production batch can be tested.

The method first developed involves X-raying the items to see if any flaws are present. When a stream of electrons strikes a metal surface in a vacuum, X-rays, which pass easily through materials but not through air, are given off. These are directed through the object, and film, sensitive to them and placed on the opposite side, will show gross, but not minor or surface, defects after being developed. Large holes inside an object show up as dark, unexposed patches on the film. In operation, electrons emitted by a heated filament are accelerated by a high direct voltage (70kV to 2MV). The higher the voltage the greater the radiation energy and the thicker the object that can be tested. As the X-ray tube is basically a diode, it is self-rectifying and a.c. can be used, simplifying the high voltage circuitry.

Surface cracks can be located (very accurately with small probes) in conducting materials by eddy-current testing. A high frequency a.c. signal is applied to one winding of the probe, inducing a current in the other winding and eddy or surface currents in the test object. The eddy currents produce a magnetic field which induces an opposing current (Lenz's law) in the second winding and thus the instrument

X-ray inspection system

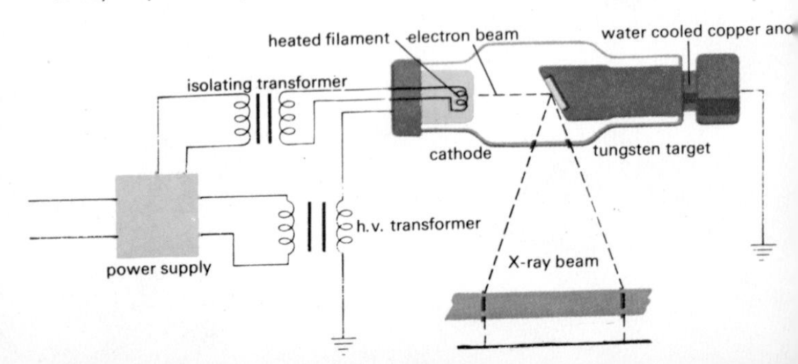

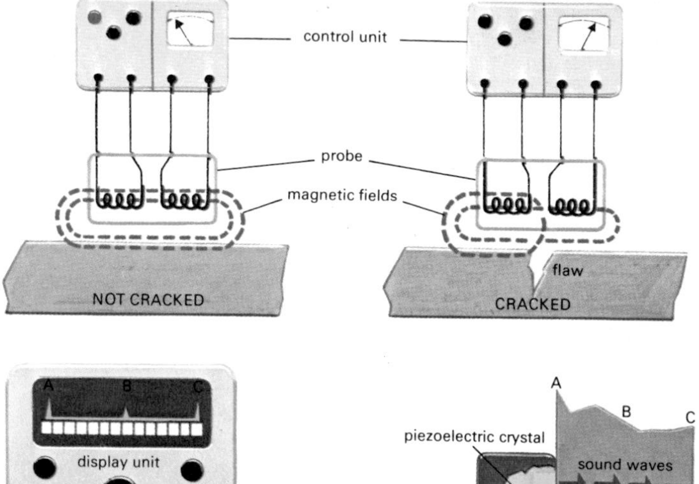

(*Top*) Eddy current testing
(*Bottom*) Ultrasonic testing

meter gives zero deflection. If a crack is present the eddy currents cannot flow, and so the meter is deflected.

The most versatile method involves ultrasonic waves. Waves of frequencies, much higher than those the human ear can detect, are generated and detected by crystals of *piezo-electric* materials. These crystals vibrate mechanically and emit high frequency waves when energized by an electric current, and they will also produce an electric current when vibrated mechanically. The waves are transmitted through the object under test and the echos are displayed on an oscilloscope. The waves will not travel through air and so they are reflected at any material/air interface. In the illustration, A represents the echo of the near-side and C the echo of the far-side of the sample. If a flaw is present, a subsidiary echo B results, the position being directly related to the depth of the deflection. As the thickness of the sample can be measured and the ratio A-B/A-C is known, the depth of the flaw can be calculated. Surface defects can also be detected by using special angle probes that direct the waves along the surface.

Current collection. (*Above*) Overhead. (*Below*) Third rail

ELECTRICITY IN TRANSPORTATION

Land transportation

Railways

The use of electricity to power locomotives is increasing rapidly. Electric locomotives have better acceleration, are faster, more reliable, require less maintenance and have a much higher power to weight ratio than do diesel powered locomotives. They require no warming-up period and will operate in a wide variety of climatic conditions.

Early systems, such as the London underground, used d.c. motors, one on each axle, supplied from mercury arc rectifier stations, situated every few miles alongside the track. A few overhead collection systems were used but the majority involved a third rail system. Speed control was obtained by means of current limiting resistors in series with the motors and by switching the motors into various series/parallel combinations.

For long distance schemes it is uneconomic to build rectifier stations every few miles so that small locomotive mounted rectifiers were adopted, the power being supplied by industrial frequency a.c., taken from the local power system. Overhead current collection techniques were adopted for safety due to the high voltages (25kV) involved.

The motors in a.c. locomotives are still d.c. energized since speed control of a.c. motors is difficult (page 111). Transformers are mounted inside the locomotive, the voltage being reduced before conversion to d.c. Mercury arc rectifiers were initially used but solid state types are now being introduced. Speed control is much easier, tappings being provided on the transformers so that the voltage can be controlled in small steps from 0 to 600 volts. Series type d.c. motors (page 110), mounted directly on the axle, are used, the body containing the transformers and rectifiers (where needed) and the control and auxiliary gear.

Electricity is also used for signalling systems and communication links. Recent developments in automatic train control, where starting, stopping and journey speed are controlled by signals picked up from electric signal sources situated alongside the track, have improved operating efficiency and safety. Speed limits and danger signals can be automatically enforced. All trains on the new London underground Victoria Line, are controlled by two men in a central control area, the train operator only opening and closing the doors and pressing a button to start the train.

An electric locomotive

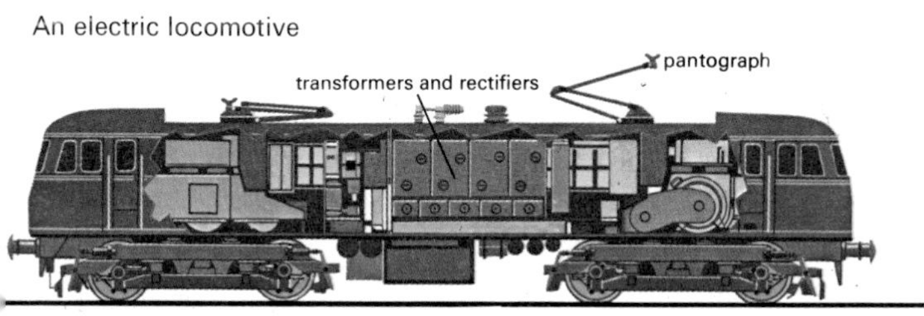

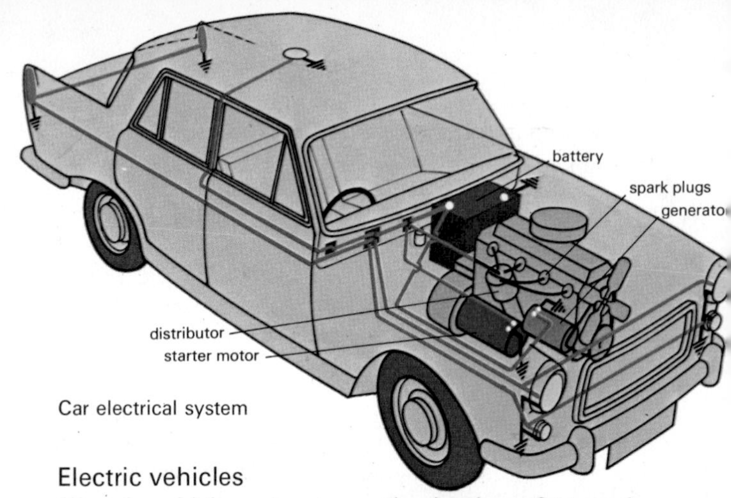

Car electrical system

Electric vehicles

Electric vehicles are not new, having been first used over a century ago and coming into real prominence about the turn of the century. At present there are over 100,000 electric vehicles in Great Britain, mainly small delivery units, and one large London store is still using vans built over fifty years ago. The general conversion from d.c. to a.c. power distribution in 1910 was a set-back, since early rectifiers were large and not suited to battery charging.

The advantages of electric vehicles are numerous, the absence of atmospheric pollution and noise being the most obvious. For 'city only' cars they are cheap to build and run. In general, maintenance is lower and life is longer (7 year average as opposed to 5 for petrol driven vehicles). The disadvantages are the short range (20 to 30 miles at 30 mph), slow speed, long charging (refuelling) times of 6 to 8 hours and the battery weight.

Present vehicles use normal lead acid batteries (page 32) since, although heavy and bulky, they are cheap and reliable with a minimum life of four years. Much research is underway to find better power sources. Fuel cells are possible, but they work best with light loads, and hybrid vehicles using small petrol driven generators have interesting possibilities.

A promising power source is the zinc-air cell, which is a combination battery and fuel cell. Energy is developed from reaction of the zinc with the oxygen in the surrounding air. Zinc is removed during discharging and deposited during charging (which only takes 2 hours). These units give six times the energy output per pound weight of bad acid types but at present only have a life of about one hundred charge-discharge cycles.

Compound motors (page 111) are used as they are more efficient, give better speed control, and allow regenerative braking (where the car wheels drive the motor as a generator, slowing the car down and giving the battery a slight charge at the same time). Speed control is normally by series resistance but electronic methods, using controlled rectifiers, are under development. One motor driving the rear wheels through the normal differential is inefficient and a separate motor on each wheel is better. This latter simplifies the utilization of four wheel drive.

Pure electric cars will probably never be high speed, long distance vehicles but, if improved batteries are developed, they will become the accepted vehicle for urban use.

Cars

Ordinary internal combustion engined vehicles also depend on electricity. The engine will not operate without electricity to supply the fuel-igniting spark. Electric motors make starting easier and keep the windscreen clear. Gauges indicate the condition of various parts of the system and electric lights provide illumination for night driving.

Experimental electric car

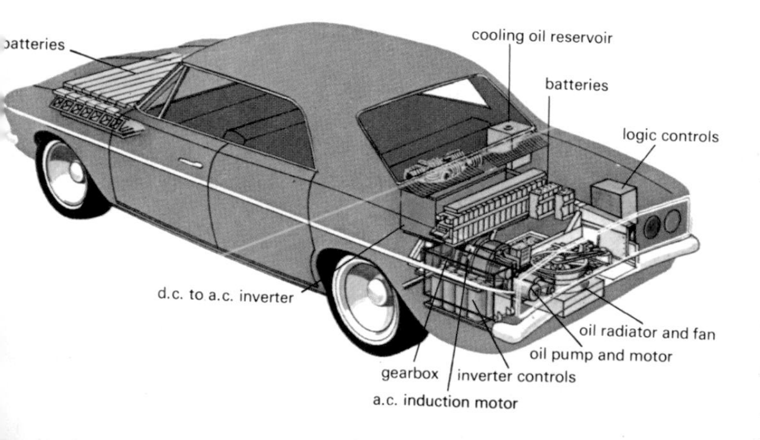

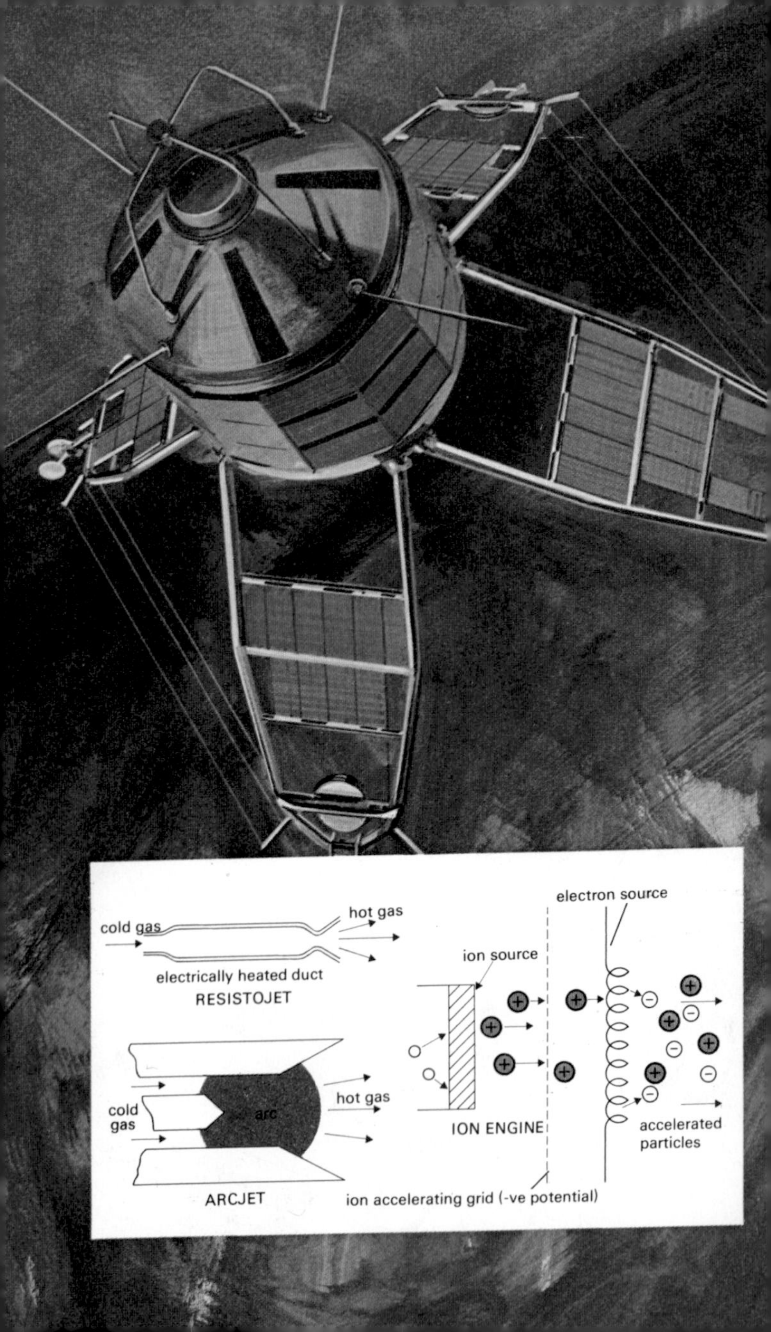
cold gas
hot gas
electrically heated duct
RESISTOJET
cold gas
arc
hot gas
ARCJET
electron source
ion source
ION ENGINE
accelerated particles
ion accelerating grid (-ve potential)

Air and space

Modern aircraft are dependent on electricity. The control surfaces move by electric motors operated by signals derived from the pilot's controls. Lighting, instrumentation, communication systems and radar all depend on electricity.

In spacecraft and satellites electricity is the only available power source. Batteries are used in conjunction with other sources that prolong their life. Solar cells (page 86) recharge batteries from the Sun's energy but, as they are not efficient converters, other sources have been developed.

Fuel cells (page 87) are the most favoured devices, having a large power to weight ratio and being proved reliable and effective in operation. They have been used in both the *Gemini* and *Apollo* space programmes, and the moon landing could not have taken place without them. Hydrogen and oxygen are used as fuel. Water is formed as a by-product, and thus no water supplies are needed and the heat generated is used to heat the spacecraft.

Electric propulsion systems are practicable in space and many research programmes are examining this aspect of their use. The simplest method is the *resistojet*, where cold gas is fed in one end of a tube (one millimetre diameter) and heated to over 1000°C, whereupon it rushes out the other end (hot gases expand enormously) with great pressure, producing a large thrust. This system is mainly used for small orbit connections of satellites. Secondly, there is the *arcjet*, which operates on the same principle but uses an electric arc to heat the gas. Consequently, higher temperatures and thus higher velocities are produced. Finally, the *ion engine* is absorbing threequarters of the American research effort in this field. Positive ions are accelerated electrostatically, very high velocities being developed, and are then ejected to give a large thrust (positive ions have much greater masses than electrons), the process being highly efficient. One drawback is that, as positively charged ions are emitted, the engine assumes a negative charge and starts attracting some of the ions back. This is prevented by injecting electrons to neutralize the propulsion beam.

Spacecraft showing panels of solar cells

Sea

Electricity has contributed much to the safety and efficiency of sea travel. An example is the lighthouse where powerful electric lights and foghorns warn ships of rock and shoal nearby. Arc lights are usually used, with a range of up to twenty miles in good weather. Electric foghorns are sometimes used but air operated ones are superior. Electric motors drive the compressors that supply the large amounts of compressed air required. Some lighthouses are completely automatic, with electronic gear to switch the lights and start the foghorn, when detectors respond to bad weather.

Lighthouses can now be made independent of electricity supplies. In Japan, a small experimental lighthouse derives its electricity from wave action. Air, compressed by the rise and fall of the waves, is directed on to turbine blades. The small generator, driven by the turbine, charges a battery and the system provides sufficient power to light a 100 watt bulb each night. On the Thames, a small lighthouse is powered by solar cells (page 86).

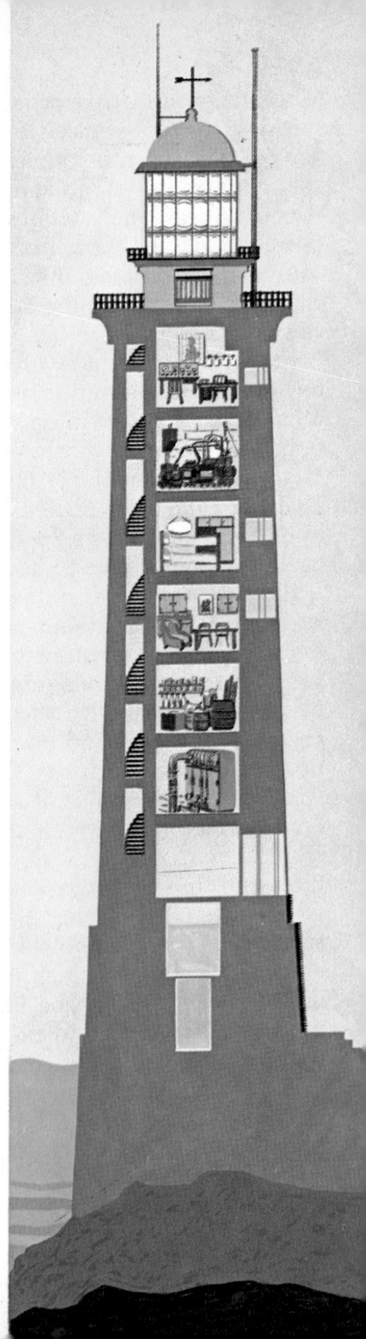

Another system, called *RIPPLE* (Radio-Isotope Powered Prolonged Life Equipment), has been developed. As Strontium 90 decays it emits neutrons which generate heat, due to collisions with other molecules, as in a nuclear reactor (page 84). This heat is converted to electricity by the thermoelectric effect (page 36). These units have a predicted life of at least five years and a number are now in service in Great Britain and Sweden.

Ships are becoming more and more dependent on electricity. Navigational gear and communication systems, internal and external, are based on it. Modern steering gear is electrically operated, as are all winches and cranes. Newer ships can almost be operated by one man, with the help of a computer, and crews have been much reduced.

Electric motors have been used as the main ship propulsion system for many years. Steam turbines operate most efficiently at speeds above 1000 rpm, but propellors only turn at about 80 rpm. Gearing is inefficient and so the propellor is operated by an electric motor, driven from a generator that is coupled to the turbine. Both a.c. and d.c. systems have been used.

RIPPLE power system. (*Left*) Buoy. (*Right*) Power unit

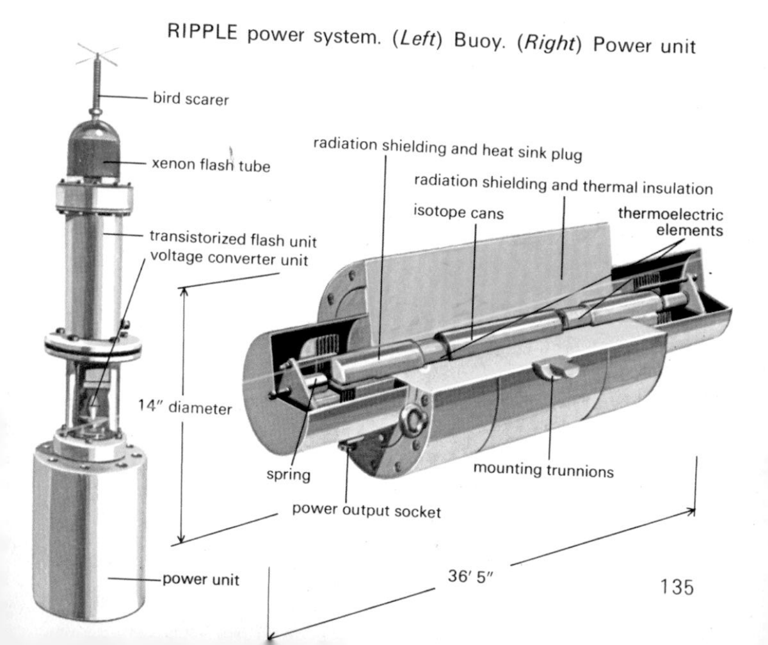

ELECTRICITY IN THE HOME

Electricity is one of the home's most versatile servants, being used for lighting, heating, cooking and entertainment. It can also be dangerous and the precautions emphasized in this section must be observed.

Electricity is normally transmitted to the house by underground *mains* cables, although in rural areas overhead lines are used. The cables are brought into a fuse box (normally 60 ampere rating), then to a watt-hour meter to measure the amount of electricity consumed, and then to the distribution board for the house wiring.

Electricity is supplied on two basic tariffs, the *standard* and the *off-peak*. Night-time consumption of electricity is about half the daytime level. The Electricity Supplier must cater for the maximum load, and so, to make full use of the

Unearthed and earthed appliances

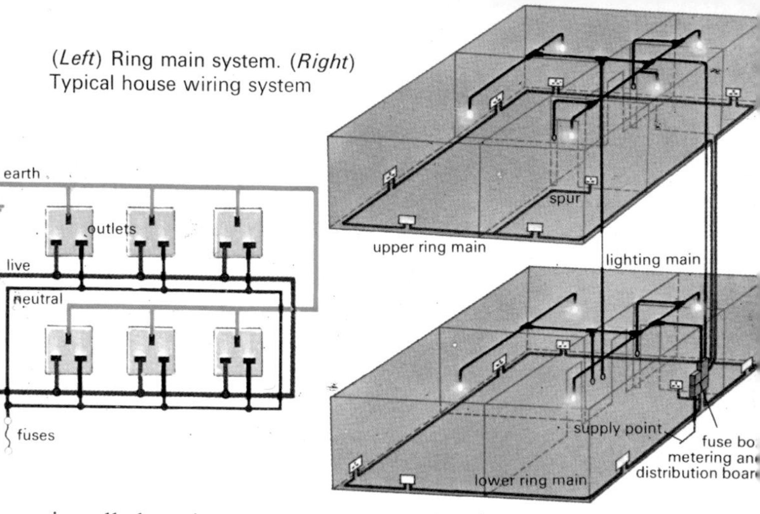

(*Left*) Ring main system. (*Right*) Typical house wiring system

installed equipment, they sell night time, or off-peak, electricity at a reduced price. The off-peak supplies are mainly used for heating purposes (page 142). Thus a separate watt-hour meter is needed for each tariff, the off-peak supply being controlled by a timing unit that switches it on in the evening and off in the morning.

Older houses are wired with a two-wire circuit comprising a *live* and a *neutral* conductor. This can be dangerous, because if the insulation of the live wire becomes damaged, the frame of the appliance becomes energized, with consequent danger of electric shock. Modern practice is to use a three-wire system, *live*, *neutral* and *earth*. Then if the live wire touches the frame, a circuit is completed via the earth lead and the fuse blows, thus removing the danger. Repeated blowing of fuses by an appliance indicates an internal fault and the appliance should be sent for repair.

The standard modern wiring system has a 30 ampere ring main with 13 ampere outlets. This means that the current is divided between two paths and, as all the sockets are unlikely to be fully loaded at the same time, smaller wires can be used (20 ampere rated wire is used for 30 ampere ring circuits). Older houses were wired on the radial system, necessitating a separate run of wires for each outlet, with consequent higher cost, due to the longer length of wire required.

One of the commonest dangers, especially in older houses, is overloading the circuits, for example by running an iron from a lamp socket. The wires are then forced to carry higher currents than they were designed for and they overheat. This can cause fires inside walls or, more commonly carpets are set alight from the wires running beneath them (which should be avoided). If there is doubt, wiring should be checked by an electrician.

All electric circuits should contain fuses, which are safety devices. These fuses interrupt the current if it exceeds a set level, thus preventing the wires overheating. A fuse consists of a strip of metal whose melting temperature is slightly above the temperature it reaches when passing the normal current. When overloaded the fuse wire temperature rises and it melts, thus creating an open circuit. The fuse must be replaced before the circuit can be used again. Early fuses were lengths of special wire usually stretched over insulated holders and secured by screws. The larger the diameter of the wire, the greater the current it will carry before melting. Modern practice is to use the cartridge fuse secured by clips. This is just a wire contained in a sand filled, insulated tube.

A recent innovation is the use of a miniature circuit breaker (MCB). The overload current causes the magnetic field in the coil to attract the plunger which, by a series of mechanical linkages, opens the contacts. When tripped, it can be reset by simply pushing a button so that a supply of spare fuses is unnecessary. The tripping current level can be easily

Overloaded circuit

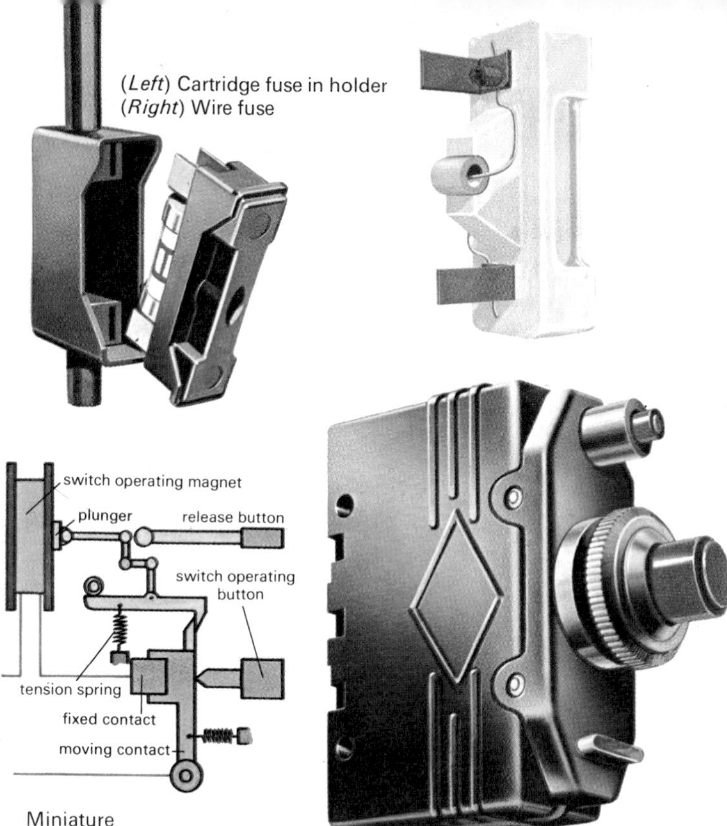

(*Left*) Cartridge fuse in holder
(*Right*) Wire fuse

Miniature circuit breaker

set by adjusting the distance the plunger must move before the MCB operates; the further the distance, the larger the magnetic field, and hence the higher the current needed.

Fuses and MCBs should be closely matched to the current being used. It is no use using a 30 ampere fuse for a lighting circuit, drawing less than half an ampere. On every piece of electrical equipment is a *nameplate*, giving its voltage, frequency, and power rating. From this the required fuse size can be determined (page 28). A general rule is to allow 4 amperes per kilowatt and then use a fuse of the next higher available rating. The preferred values are 5 and 15 amperes for wire and 3 and 13 amperes for cartridge types.

Plugs must be used carefully, especially in older establishments. Multitudes of plugs, unfused and normally rated at 2, 5 or 15 amperes, were, and still are, available. If a 5 ampere plug were used on a 3 kilowatt electric fire, drawing 12 amperes, it would overheat and possibly burst into flame. The standard modern plug carries 13 amperes and has an internal fuse holder, which means that it can be used for every requirement by changing the fuse.

Adaptors are available to turn a single outlet into a multiple one but these must

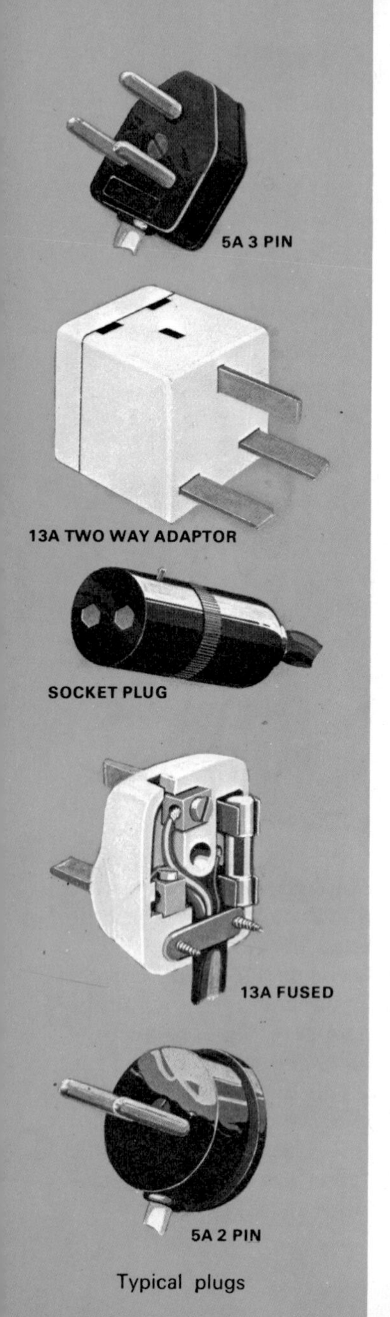

Typical plugs

(*Left*) Typical plugs
(*Below*) Wire identification.
1-neutral, 2-earth, 3-live

INTERNATIONAL STANDARD 1 2 3

BRITISH STANDARD 1 2 3

(*Below*) Misuse of plugs

also be used with caution. Just because one outlet has been changed to three by an adaptor, it does not mean that three times the current can be used, since the supply wire is still the same size! Two extra outlets per socket is the maximum that should be used, not 5 or 6 as is very common. If a number of appliances must be run simultaneously, further outlets wired back to the main distribution board should be installed.

Plugs should fit tightly into outlets, otherwise arcing and consequent overheating result, with the danger of fire.

When plugs are being attached to apparatus, the wires, which are live, neutral and earth, must be connected to the proper terminals and these are colour coded for this purpose. The British Standard and the International Standard wire colour codes are illustrated. Both are now in use but after 1971 the latter only were used for new equipment in Britain. On two wire appliances the earth is omitted and either wire may be connected to the live, the other then going to the neutral.

Bare or loose wires should be replaced immediately. Older style sockets can be dangerous to small children, since fingers or small metal objects can be inserted. Modern 13 ampere sockets of reputable manufacture have sliding protective covers that only open if the correct plug is used. Burns, shocks and even death can be obtained from normal 240 volt mains. The danger is increased in wet areas, such as bathrooms, where the water can act as a conducting path to an earthed metal drain pipe.

Water should never be used on fires caused by electricity. Normal tap water is conducting and it is possible to electrocute one's self. The electricity supply should be switched off, if possible, and the fire smothered, if small. Otherwise a fire extinguisher, especially designed for electrical fires, should be used. If someone receives an electric shock, they should not be touched with bare hands, because the helper may also receive a shock. The supply should be switched off first, or the person dragged away with an insulated object, such as a wooden chair. Electric shock normally paralyses the nervous system controlling breathing and the victim suffocates, and so artificial respiration should be started at once.

Apart from lighting, the major use of electricity in the home is for heating. All electric heaters consist, basically, of a metal element of fairly high resistance wire, about 60 ohms per kilowatt rating. As shown previously, the heat produced is proportional to the product of the current squared and the resistance of the wire. The element wire is usually composed of nichrome (an alloy of nickel and chromium), which has a very high melting temperature so that it is not damaged at the temperatures required to produce the heat.

The commonest type of electric fire is an assembly of one or more elements, usually of one kilowatt rating, in a frame with a polished reflector to beam the heat into the room.

The radiant heater, has a similar construction, but the elements are composed of *inconel*, which gives off invisible infra-red radiation. This warms objects that it touches, thus being only suitable for personal heat.

Fan heaters have a small fan circulating air over the element, and ejecting the warm air to heat large areas.

Convector heaters utilize

Convector heater

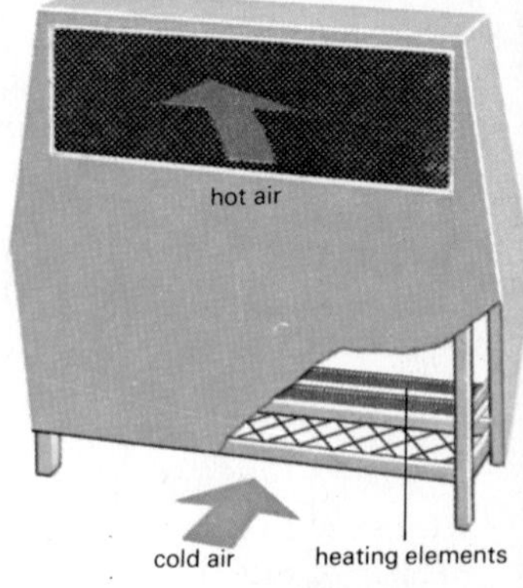

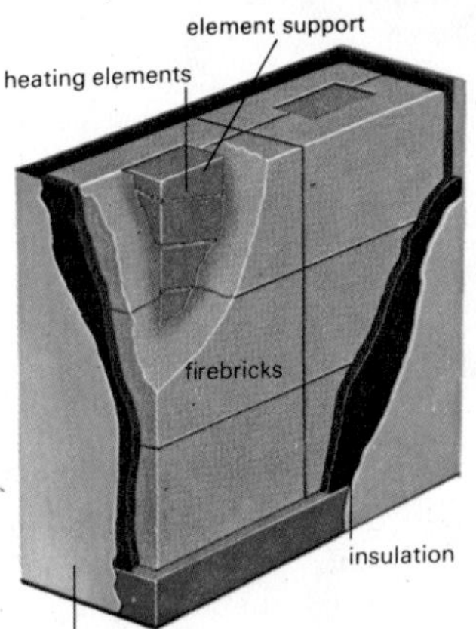

Storage heater

Electric fire

the fact that hot air rises. The element is contained in an upright case with openings top and bottom. The rising warm air flows out into the room, sucking cold air in at the bottom. Again large areas are heated but without the draughts that fan heaters create.

To make use of the cheaper off-peak electricity tariff, heaters have been developed which *store* heat during the night and release it during the day. The *free standing* type of storage heater consists of heating elements embedded in *fire-bricks*, which are heated to a dull red during the night. As they cool during the day the heat is given off to the surroundings. In the *underfloor* system, the sand cement mixture stores the heat, and as the whole floor area is available the elements do not need to be heated to very high temperatures.

Electricair central heating is based on the storage heater principle. A large central unit has a fan mounted inside it and the hot air is distributed to the various rooms via ducts. A thermostat is incorporated to keep the room temperatures constant, the fan only being switched on when necessary.

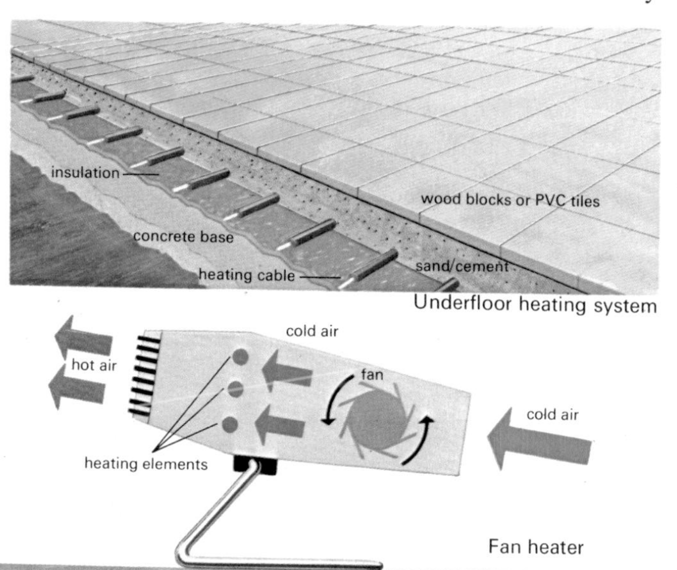

Underfloor heating system

Fan heater

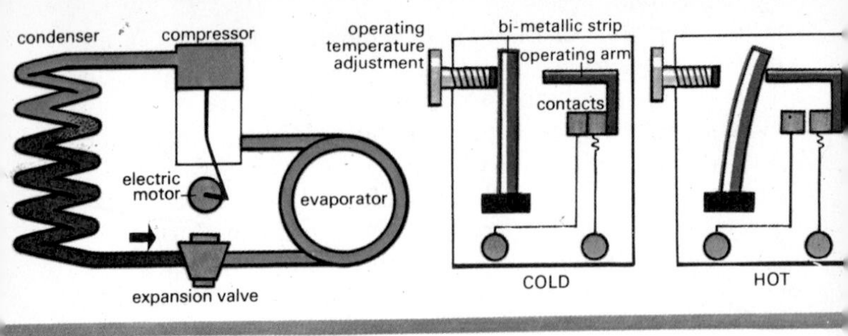

(*Left*) Refrigeration cycle. (*Right*) Thermostat

The many electrical household appliances now in use may appear complex but are, in fact, only applications of the basic principles of electricity, heat and electromagnetism.

The two most important appliances are probably the cooker and the refrigerator. The cooker just makes use of the heating effect of an electric current. Sophistication is added with devices such as temperature controllers (thermostats) and timers that pre-select the switching on time and also the duration of cooking. Thermostats keep the temperature at a pre-set level by switching off the current when the required temperature is reached and switching it on again when it falls below this level. They consist of a set of contacts, which carry the heating current, operated by a bi-metallic strip (two different metals bonded together). This strip bends as the temperature increases, due to the different expansion rates of the two metals, and the contacts open, closing as the temperature decreases. By mechanically adjusting the amount of bending required to open the contacts, the operating temperature of the oven can be varied over a wide range. A recent development has been the *self cleaning* oven. A special control is incorporated, which locks

the oven door and raises the temperature to 480°C (900°F), burning any grease and food soils and leaving a small amount of fine ash which is easily dusted away. A special filter in the oven vent renders the smoke colourless and virtually odourless.

Refrigerators, although powered by electricity, depend on the fact that when vapourized a fluid becomes cold and removes heat from its surroundings, and that this heat may be given off when it is compressed back into fluid form. In an electric refrigerator an electric motor drives the pump which compresses and circulates the fluid that must have a low boiling point. The pressure of the fluid is reduced by an expansion valve so that it vaporizes in an evaporator, which is inside the area that is required to be cooled. In the condenser, it is then compressed (compressed vapour liquifies at a low temperature) outside the cold area, giving up heat which is dissipated into the room.

Electric clocks are just small synchronous motors, whose speed depends on the supply frequency. As this is kept constant by the supply authorities, these clocks keep good time. Digital clocks, with numbers indicating the time instead of hands, have a mechanism that counts the number of revolutions of the motor and changes the displayed number as required.

Electric kettles just contain heating elements, whereas irons, toasters and immersion heaters have heating elements with built-in thermostats for temperature control. In toasters, the thermostat also operates the mechanism which causes the toast to *pop-up* when the time is up. Mixers, carvers and toothbrushes are basically electric motors.

Other appliances doubtless spring to mind but with a little thought the basic operating principles should be quite obvious.

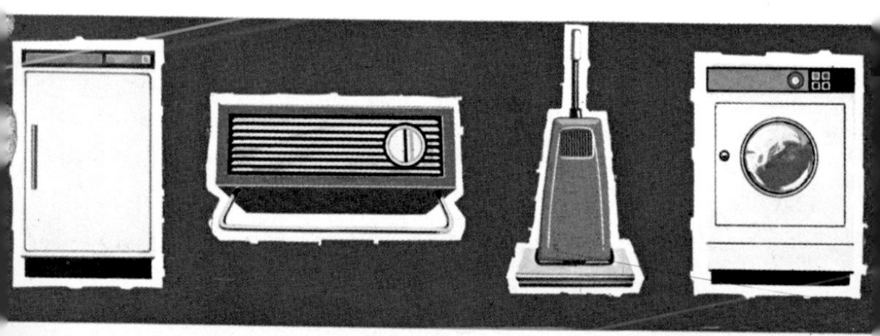

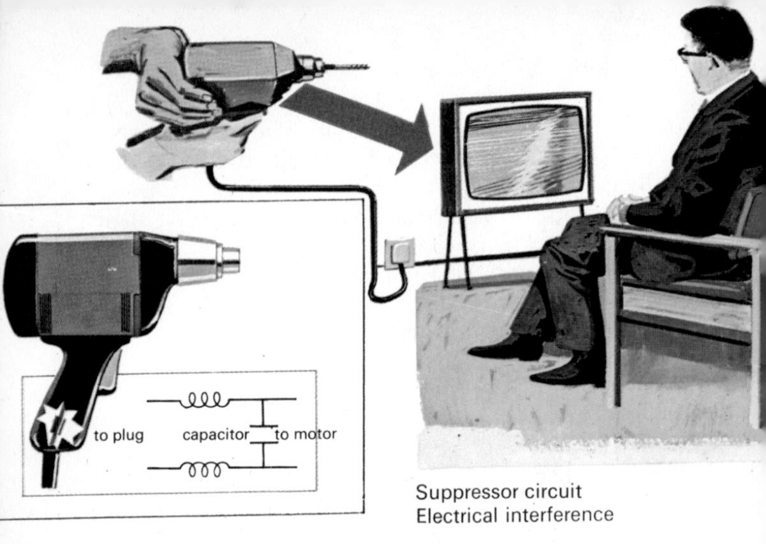

Suppressor circuit
Electrical interference

Electrical appliances can interfere with television and radio receivers. This interference appears as crackling and loud bursts of noise in radios and as snow or lines on the television screen. The causes are sparks inside an appliance, resulting from bad connections or the type of motor used. A majority of the motors used are of the universal type (page 111) which spark every time the brushes move from one commutator segment to the next. As these sparks are of very short duration, very high frequencies are present which radiate into the surroundings. In fact, the first wireless communication, using morse code, relied on sparks to generate the radio waves for transmission.

Interference may be propagated by three distinct means; by *direct radiation* (try holding an electric shaver near a radio), which is limited to a few yards range; by *conducted interference*, which travels along the mains lead into the power line so that one faulty appliance can affect houses over a large area; and as *interference from the mains* which has been injected into the mains at one point and is being radiated from it at another into equipment, such as a transistor radio, which is not connected to the mains.

Other causes of interference are atmospheric disturbance (lightning and electrical storms), overhead power lines (very rarely, unless directly underneath, as they are designed to be interference free) and automobile and motorcycle ignition systems. The latter are particularly prone to affecting television sets at ranges up to ten yards.

The only method of stopping or *suppressing* interference is to cure it at the source. Interference can travel through many unexpected paths, including earth wires which appear to be solidly earthed. Faulty connections are obviously cured by remaking them properly. Motor interference is suppressed by filter circuits, called *suppressors*, being fitted as close to the motor as possible. The capacitor values are chosen to present a high impedence to the 50Hz supply but a low one to the high frequencies, and conversely the chokes are chosen to present a low impedence to the 50Hz but a high one to the high frequencies. Any high frequencies then circulate around the motor and the capacitor and cannot pass into the mains supply system.

All modern appliances are required by law to have suppressors fitted. Older appliances should have them fitted as it is an offence to interfere with television and radio. The British General Post Office has a section, equipped with special instruments, which is constantly engaged in tracking down interference, both industrially and in the home.

ELECTRICITY IN THE FUTURE

So far we have considered the history and present development of electricity and now we will look at things to come. Some of these are purely speculative, some are new applications of existing knowledge, others are having considerable amounts of money spent on their development, whilst a few are in use, although generally only on an experimental basis.

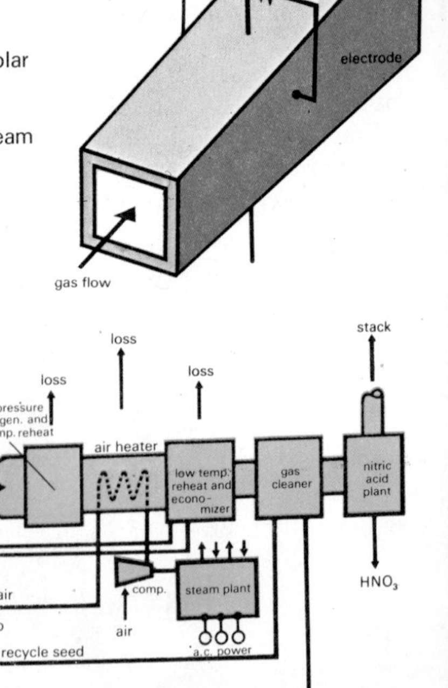

(*Above right*) House with solar cells on roof. (*Right*) Magneto-hydrodynamics. (*Below*) Diagram of MHD steam plant

Generation

Better and more efficient methods of producing electricity in large amounts are continually being sought. The most promising field at the moment is *magneto-hydrodynamics*, abbreviated to MHD.

Power output from a generator is directly proportional to the velocity at which the lines of magnetic flux are cut. Conventional generators have conductors rotating through a magnetic field, or vice versa, but speeds above 500 feet per second cannot be obtained, due to mechanical limitations. In an MHD set, conducting gas moves through a magnetic field at velocities up to ten times greater, with consequent increased power output.

This stream of very hot conducting gas passes through a duct having a perpendicular magnetic field and electrodes along it. The gas (air with added oxygen) is heated to 3000°C by conventional means and is then made conducting by *seeding* with sodium, which vaporizes into conducting ions in the hot gas.

These systems produce direct current and are thermally very efficient. They are also robust, having no moving parts. The hot gases can afterwards be used to heat boilers to produce electricity by conventional *thermal* means. Thus the power output of an existing thermal power station could be markedly increased for a capital outlay but no increased

Nuclear fusion

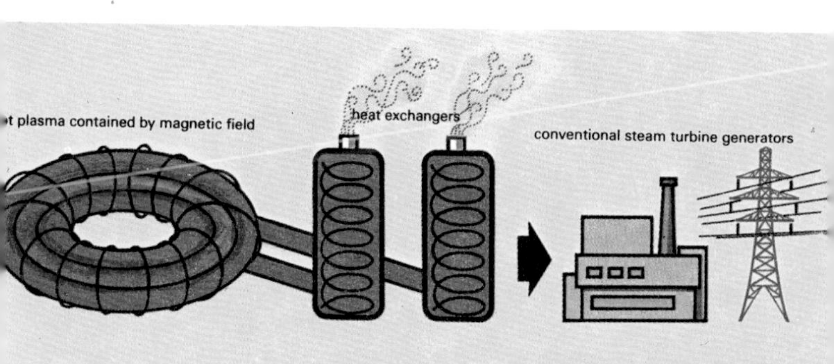

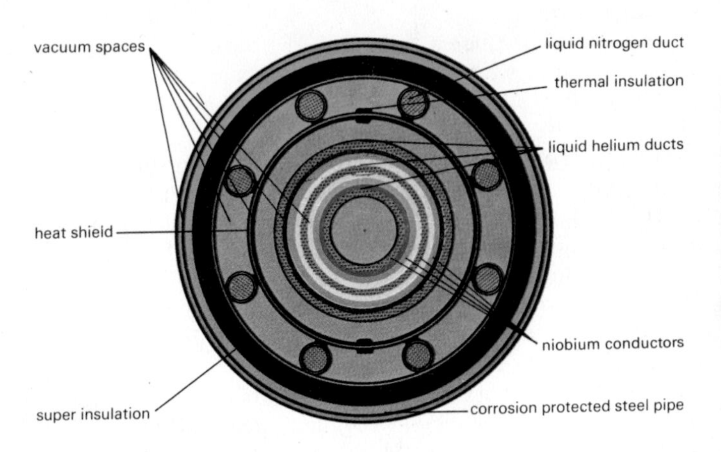

Superconducting cable

running costs. Laboratory systems have been built but no practical system is envisaged, at present.

Another system for producing electricity that is being investigated at present is *nuclear fusion*. Existing nuclear plants work on the fission principle, which is the breaking up of atoms to release energy. The fusion process gains even larger amounts of energy from the joining together of atoms. A system would start by heating hydrogen gas, by a burst of current, to the temperature where fusion begins. Once this is attained the process is self-sustaining. The resultant plasma (hot ionized gases at 100 million °C, hotter than the sun) is contained by a magnetic field since no known material will stand these temperatures. The heat would be used, with lithium beryllium salts as the transfer medium, to produce steam to drive conventional turbines. At present, the biggest problem is containing the hot gases inside the magnetic field, since not enough fundamental knowledge of plasma behaviour is available. Probably by the year 2000, this type of power source will be operational.

Another possible development would be to make all houses electrically self sufficient by having their own generators. In sunny climates this could be achieved by means of solar cells on the roof, and, in duller climates, by the use of fuel cells or nuclear powered thermoelectric devices. The efficiencies of all these generators must be markedly improved before this is possible, but a few small installations are now in operation, for example in Switzerland a remote relay television transmitter is powered by a fuel cell which needs refuelling once a year.

Transmission

The next step in electric power transmission will be to utilize the phenomenon known as *superconductivity*. In this state, conductors have zero resistance and large currents can thus flow in very small wires. Superconductivity only occurs at very low temperatures, close to absolute zero (– 273°C), where all molecular motion ceases. As the electrons are orbiting very slowly around the nucleus, they can easily be detached and thus large numbers of electrons are available and a very small voltage causes large current flows. In theory, the current flow should continue indefinitely, since there is no resistance to flow, but in practice there is some residual resistance and the currents eventually die away if voltage is not continuously applied.

Microwave power transmission

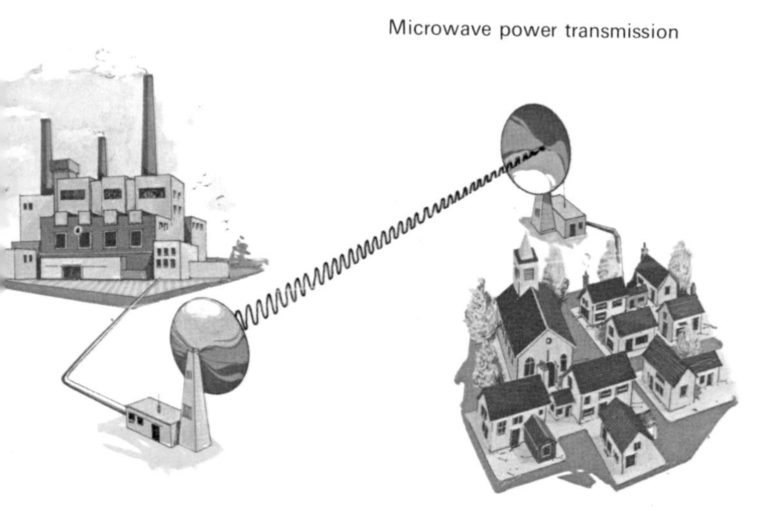

Most metals are superconducting at −270°C, which is the temperature of liquid helium. The transition to a superconducting state as the conductor is cooled is verv abrupt. A full explanation is beyond the scope of this book.

Liquid helium is used to cool the conductor in a proposed superconducting cable. The greatest problem is preventing external heat getting into the system. Vacuum is used both as electrical and thermal insulation and liquid nitrogen at −200°C as an outer heat shield (liquid nitrogen is much, much cheaper than liquid helium). Very thin conductors, 0·0025 centimetres thick, are used with current densities of 15,000 amperes per square centimetre. Thus only low voltages are required to transmit high powers. Laboratory experiments and design studies have shown this system to be practicable and economic.

As magnetic fields depend on the amount of current flowing, superconducting coils can produce extremely large fields for very little power input. Motors using superconducting coils to produce the magnetic fields have been constructed, and one is being installed, to operate a cooling water-pump, in service trials at Fawley Power Station.

The major drawback to superconductivity is the large amount of power required to operate the refrigeration plant that produces the liquid helium. Proposed systems must be carefully considered and designed so that the increased power transmitted is not all used to provide for the needs of the refrigeration plant.

Microwave power transmission is another possibility. Microwaves are very high frequency signals (1000 million Hz) which travel along the surface of a conductor, not inside it. These are at present used for communication systems, such as satellite transmission, where the power requirements are small. In theory a 4 foot diameter tube could, using microwaves, carry all the power transmitted by two 400kV overhead lines, but as yet no practical means of converting high powers from microwaves to usable lower frequencies (50Hz), or vice versa, have been devised. It should become possible to broadcast power to homes and vehicles. A small experimental model helicopter has been kept hovering by microwave power transmitted from the ground.

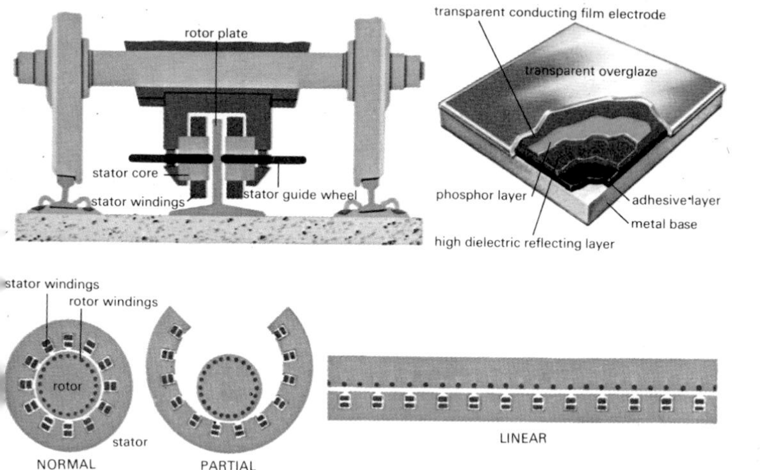

(*Top left*) Railway usage of linear motor. (*Top right*) Electro-luminescence. (*Bottom*) Imaginary evolution of linear motor

Transportation

The use of electricity in transportation is bound to increase. Electric cars will be developed to a state of high efficiency, necessitated by the anti-pollution laws now being promulgated. To this end, more efficient and lighter batteries are being investigated. Numerous experimental cars, for both town and country driving, are on the road and doubltless full production is not far off.

Another device, that has been known for years but is only now being seriously studied, is the linear motor. One of these was used twenty-five years ago in an experimental catapult for launching jet aircraft.

A linear motor is just a normal motor unrolled, the motion thus becoming linear instead of rotary. In practice the rotor is usually the fixed portion, being just a sheet of conducting material, whilst the moving stator has normal windings attached.

A possible use for this motor is for propelling a train. As the traction effort is independent of contact between the wheels and the track, the overall efficiency of the vehicle is much higher. Further developments would remove the wheels and use the *air cushion* principle to create an electrically powered, tracked hovercraft, which would use the rotor plate as a guide-rail. This system would have virtually no friction losses, very high speed, and quiet transport. At present, a number of countries (including the United Kingdom) are investigating the system.

Electromagnetic levitation has also been proposed for rapid transport systems and the development of superconducting magnets makes this a feasible suggestion.

Illumination

Certain crystals (usually zinc sulphide) glow when placed between two a.c. energized electrodes. This is a *cold* light, giving off no heat, and is therefore very useful. Lamps can be made in any shape, for example in mosaic designs on walls or ceilings. At present, the process is not very efficient and so it is finding its main use in illuminated signs. The light output increases as the frequency of the energizing voltage is raised, up to about 2000Hz. Above this dielectric heating reduces the output. This phenomenon is made use of in aircraft for *No Smoking* and *Fasten Seat Belt* signs, since the apparatus is light and occupies little space (0·25 inches), the letters are completely invisible when unenergized, and the aircraft 400Hz electricity supply gives increased light output that is adequate for daylight use.

CONCLUSIONS

It has been shown that electricity is the flow of fundamental particles of matter (electrons) and that use of it is based on four effects resulting from the flow. These four effects are listed below.

(1) Heating effects result from the collisions of the moving electrons with the stationary atoms.

(2) Magnetic effects result from the setting up of magnetic fields by the electron motion.

(3) Chemical effects result from interaction of the electrons with atoms of the other materials involved.

(4) Charge storage effects result from deposition or removal of electrons.

The reader should now have a basic understanding of these effects and how they are applied in everyday life and also have an idea of future trends. Many facets have had to be omitted and others have been only briefly mentioned, but this is unavoidable in a book of this size. On the following page is a list of selected references for those who wish to learn about the subject in more detail, and these will in turn mention other sources for deeper delving.

The scope of electrical applications is not static. It is continually changing and progressing at a rapid pace. Last year's laboratory curiosity is this year in common use and by next year will be superceded by even newer developments. Some of the uses of electricity mentioned as being under development will most likely be in use by the time this book is published. However, the fundamental properties of electricity do not change and any new device can be understood if it is analyzed in terms of its basic constituents.

Looking ahead into the realms of conjecture, the next great step will probably be the development of cheap, simple and light electric propulsion systems for spacecraft, which will make spaceflight as common as flying is today.

Electricity is a good servant but a hard master, being especially dangerous because it gives no physical indication of its presence. Injury and even death can result from even the lowest voltages and so in all practical dealings the outlook should be *safety first*.

BOOKS TO READ

For general introductions to electricity and related subjects, the following books are recommended and should be available through bookshops and/or public libraries.

Atomic Energy by Matthew J. Gaines. Hamlyn, London, 1969.

Basic Electronics for Scientists by J. J. Brophy. McGraw-Hill, New York, 1966.

Computers at Work by John O. E. Clark. Hamlyn, London, 1969.

Electrical Engineers' Reference Book (12th edition) by M. G. Say (Editor). Newnes, London, 1968.

Electronics by Roland Worcester. Hamlyn, London, 1969.

Exploring Electricity by H. H. Skilling. The Ronald Press Company, New York, 1948.

Fundamentals of Electricity (5th edition) by K. C. Graham. Technical Press, London, 1968.

Fundamentals of Electricity and Magnetism by L. B. Loeb. Dover Publications Ltd, New York, 1961.

Principles of Electrical Technology by H. Cotton. Pitman and Sons, London, 1967.

The Universal Encyclopedia of Machines (*or how things work*). George, Allen and Unwin, London, 1967.

INDEX

SOME OTHER TITLES IN THIS SERIES

Arts
Antique Furniture/Architecture/Art Nouveau for Collectors/Clocks and Watches/Glass for Collectors/Jewellery/Musical Instruments/ Porcelain/Pottery/Silver for Collectors/Victoriana

Domestic Animals and Pets
Budgerigars/Cats/Dog Care/Dogs/Horses and Ponies/Pet Birds/Pets for Children/Tropical Freshwater Aquaria/Tropical Marine Aquaria

Domestic Science
Flower Arranging

Gardening
Chrysanthemums/Garden Flowers/Garden Shrubs/House Plants/ Plants for Small Gardens/Roses

General Information
Aircraft/Arms and Armour/Coins and Medals/Espionage/Flags/ Fortune Telling/Freshwater Fishing/Guns/Military Uniforms/Motor Boats and Boating/National Costumes of the world/Orders and Decorations/Rockets and Missiles/Sailing/Sailing Ships and Sailing Craft/Sea Fishing/Trains/Veteran and Vintage Cars/Warships

History and Mythology
Age of Shakespeare/Archaeology/Discovery of: Africa/The American West/Australia/Japan/North America/South America/Great Land Battles/Great Naval Battles/Myths and Legends of: Africa/Ancient Egypt/Ancient Greece/Ancient Rome/India/The South Seas/ Witchcraft and Black Magic

Natural History
The Animal Kingdom/Animals of Australia and New Zealand/ Animals of Southern Asia/Bird Behaviour/Birds of Prey/Butterflies/ Evolution of Life/Fishes of the world/Fossil Man/A Guide to the Seashore/Life in the Sea/Mammals of the world/Monkeys and Apes/Natural History Collecting/The Plant Kingdom/Prehistoric Animals/Seabirds/Seashells/Snakes of the world/Trees of the world/Tropical Birds/Wild Cats

Popular Science
Astronomy/Atomic Energy/Chemistry/Computers at Work/The Earth/Electricity/Electronics/Exploring the Planets/Heredity/ The Human Body/Mathematics/Microscopes and Microscopic Life/ Physics/Psychology/Undersea Exploration/The Weather Guide